Molecular Insights into Development in Humans

Studies in Normal Development and Birth Defects

Molecular Insights into Development in Humans

Studies in Normal Development and Birth Defects

Moyra Smith

University of California, Irvine, USA

World Scientific

NEW JERSEY • LONDON • SINGAPORE • BEIJING • SHANGHAI • HONG KONG • TAIPEI • CHENNAI

Published by

World Scientific Publishing Co. Pte. Ltd.

5 Toh Tuck Link, Singapore 596224

USA office: 27 Warren Street, Suite 401-402, Hackensack, NJ 07601

UK office: 57 Shelton Street, Covent Garden, London WC2H 9HE

Library of Congress Cataloging-in-Publication Data
Smith, Moyra, author.
  Molecular insights into development in humans : studies in normal development and birth
defects / by Moyra Smith.
    p. ; cm.
  Includes bibliographical references and index.
  ISBN 978-9814630580 (hardcover : alk. paper)
  I. Title.
  [DNLM: 1. Human Development. 2. Congenital Abnormalities--genetics. WS 103]
  RB155.5
  616'.042--dc23
                                        2014041575

British Library Cataloguing-in-Publication Data
A catalogue record for this book is available from the British Library.

Typeset by Stallion Press
Email: enquiries@stallionpress.com

Printed in Singapore

This work is dedicated to four friends and colleagues
on three continents who have encouraged and inspired me,
Dr. Susan Dyson and Dr. Simon Prinloo in South Africa,
Dr. David Hopkinson in England, and Pamela Flodman in the USA.
Simon, thank you for being a catalyst.

PREFACE

"And meanwhile the most incredible miracles are happening all around us, stones when we lift them and let them go fall to the ground, the sun shines, bees visit flowers, seeds grow into plants, a cell in nine months multiplies its weight a few thousands and thousands of times and is a child; and men think creating the world they live in".

Aldous Huxley, 1924

And the miracle extends beyond multiplication of cells to their differentiation to multiple different tissues and specialized organs. It is the potential for pluripotency and mechanisms of differentiation that particularly inspire further study.

Our concepts of pluripotency and reversal of differentiation were greatly expanded with reports in 2006 that the introduction into the cell of only four transcription factors could promote differentiated cells to become pluripotent. Intense studies have been carried out over the past decade to determine the factors required to promote differentiation of pluripotent cells to differentiated cells. These studies have relevance not only to regenerative medicine but also enhance our concepts of normal development.

In this book, along with descriptions of development of particular organ systems, I explore aspects of relevant studies on pluripotent stem cells.

A second quotation relevant to the goals of this book and to studies on congenital malformations comes from Dr. James Paget in 1882.

"We ought not to set them aside with idle thoughts or idle words about "curiosities" or "chances". Not one of them is without meaning; not one that might not become the beginning of excellent knowledge, if only we could answer the question-why is it rare? Or being rare why did in this instance happen".

Dr. David Smith used this quotation in 1970 in his seminal publication, *Recognizable patterns of human malformation*. He inspired generations of clinicians and researchers to carry out careful examinations, to document patterns of human malformation, to document family history and to track the long-term consequences of malformations.

Availability of techniques for detailed chromosome analyses in the last three decades of the 20th century revealed that a number of recognizable syndromic forms of human malformation were due to trisomy or monosomy of specific chromosomes or dosage changes of specific chromosome segments.

The availability of human DNA sequence data for each of the human chromosomes through efforts of the Human Genome Project and enhanced capabilities of sequencing DNA from individuals has led to identification of specific gene mutations that give rise to structural or functional defects. In addition scientists are now able to explore the epigenome, secondary modifications of DNA, histone and chromatin and nucleosome positioning that play critical roles in regulation of gene expression.

In this book, I present examples of instances where insights into steps in embryonic development have been gained through discovery of specific gene alterations or mutations in cases with developmental defects.

CONTENTS

CHAPTER 1

GENOMES, GENES, STRUCTURE AND FUNCTION

Molecular embryology involves analysis of gene expression and delineation of the key gene products involved in determining differentiation at
the cellular and tissue level.

Overlapping Layers within Genome Architecture

Overlapping layers within the genome architecture need to be taken into
account in analyzing gene expression, these include:

(a) The linear sequence of DNA and linear gene structure;
(b) The embedding of DNA in histone rich chromatin that undergoes
    modifications that impact gene expression;
(c) The arrangement of DNA and chromatin within the nucleus at differ-
    ent stages of the cell cycle.

During metaphase distinct chromosomes can be identified. More precise identification of each of the human chromosomes was initially
achieved through development of specific histological staining techniques
that revealed banding patterns on chromosomes (Caspersson *et al.*, 1970;
Seabright, 1972). Molecular cytogenetics began in 1977 when labeled
DNA probes were hybridized to chromosomes.

Analysis of folding and looping of DNA and chromatin in interphase
nuclei and of three-dimensional structure is currently being actively analyzed. There is evidence that gene regulation is influenced by looping of
chromatin and DNA and by dynamic changes in the positioning of loops
on which specific genes and regulators are positioned. Fluorescence
in-situ hybridization (FISH) techniques provided means to examine the
location of specific gene targets within the nucleus. FISH studies initially provided evidence that specific chromosomes have preferred
positions in the nucleus. The arrangement differed in different cell types
(Bickmore, 2013).

Molecular Organization of Eukaryotic Genes

The molecular organization of eukaryotic genes, including human genes has been actively ongoing since the late 1970s when it became clear that these genes contain coding segments (exons) interspersed with non-coding segments introns (Leder, 1978). Analyses revealed that transcription is initiated from the 5 prime (5′) end of the gene from a site adjacent to promoter sequences that are located further 5′ (upstream) of the transcription initiation site. The promoter sequences are not transcribed but comprise elements that facilitate transcription: specific regions within the promoter bind polymerases essential for transcription.

Detailed molecular studies revealed that the primary mRNA transcript undergoes cleavage at specific points and subsequent rejoining and that this splicing process removes introns. The precise location of splicing sometimes varied depending on the nucleotide sequences present at specific positions. Specific nucleotide sequences at the 3′ end of the transcript directs binding of an endonuclease that lead to cleavage at the 3′ end and this cleavage site then polyadenylated.

Developmental Control of Gene Expression

Development is characterized by temporal and spatial differences in gene expression. Development requires the integrative action of gene promoters and cis-regulatory elements including those that lie close to promoters and those that lie at some distance from transcription start sites. Important cis-acting elements that impact gene expression include enhancers, silencers, insulators and transcription factor binding sites.

In this section, I review information on promoters, on cis-regulatory elements including enhancers and transcription factors and their binding sites. Following this aspects of transcription, alternative splicing, alternate generation of 3′ end sequences and aspects of translation are then reviewed.

Pal *et al.* (2011) emphasized that analysis of transcription and its regulation is essential for deciphering cell and tissue specific gene functions. Alternative transcription initiation and alternative transcription termination give rise to alternative mRNA transcripts and these may undergo alternative splicing and then give rise to alternate protein isoforms.

Promoters

Promoters are not transcribed; they contain sequence elements that enhance transcription capability. Promoters include core domains and regulatory domains. Key sequence elements within the promoters include the polymerase II binding site and transcription binding sites. Protein coding genes are transcribed by RNA polymerase II and core elements included in these genes include TATA boxes (sequence rich in thymine and adenine), CpG islands (repeats of cytosine and guanine) are often present in the promoter regions. Chromatin modifications are key in determining transcription initiation and are discussed in the subsequent chapter on epigenetics. TATA boxes are present in promoters of approximately 24% of human genes. Key to transcription of TATA box-containing promoters is a TATA box binding protein (TBP). This protein binds to the TATA box and acts to properly position the RNA polymerase II. It also serves as a scaffold for the assembly of transcription factors.

Alternate promoter use

Different stages of development and different cellular and tissue conditions may require use of different promoters that are differentially regulated. Furthermore sequence differences in promoters lead to binding of different transcription factors.

Factors that determine the use of alternate promoters include DNA sequences in regulatory regions and histone modifications. Pal *et al.* (2011) reported that 50% of the multi-transcript genes they studied used multiple promoters. They also determined that genes which used multiple promoters also exhibited alternative transcription termination and alternative splicing. They carried out mRNA sequencing and epigenetic analysis to develop an inventory of transcript variants that occur in development of the cerebellum in the mouse. Their studies revealed extensive changes in the expression of transcript variants of a number of different genes during differentiation of the granular cells of the cerebellum.

Examples of genes with multiple promoters

The uridine diphosphate glucuronosyl transferase 1A (*UGT1A*) gene on human chromosome 2q37 has 13 alternative promoters. The *BDNF* gene that encodes "brain derived neurotrophic factor" is expressed from nine different promoters. The gene gives rise to multiple different transcripts that result from use of different promoters, different transcription start sites and from alternative splicing of exons. Neuronal activity impacts *BDNF* transcription and splicing (Autry and Monteggia, 2012).

Sakata *et al.* (2009) reported that *BDNF* promoter IV plays a particularly important role in *BDNF* transcription. They reported that absence of that promoter in mice led to deficits in specific neurotransmitter functions in gamma amino butyric acid (GABA) ergic interneurons in the prefrontal cortex.

Methods to analyze alternate promoters of a specific gene:
CAP sequencing

Messenger RNA contains a 5′ CAP sequence in which 7-methylguanosine is linked to the first transcribed nucleotide in a phosphodiesterase bond. The CAP synthesizing complex is associated with RNA polymerase. CAGE sequencing is a sequencing method that involves biotinylation of this 5′ 7-methylguanosine and then selection through streptavidin binding. This capture enables sequencing of 5′ mRNA adjacent to the 5′ CAP. CAGE sequencing has revealed that most genes have multiple promoters and there is frequent tissue specific or developmental stage specific promoter usage. Faulkner *et al.* (2009) reported that repetitive DNA and retro-transposons (ancient sequences that can amplify themselves and are mobile in the genome) are often associated with the 5′ region of protein coding genes and may serve as alternate promoters. They proposed that retro-transposon transcription has a key influence on the transcriptional output of the mammalian genome.

Promoters have a 3′ splice donor but lack a 5′ splice sequence. Each different promoter can then bind to the downstream exon of a gene through binding to the splice acceptor adjacent to that exon.

Transcription

Transcription factors are key constituents of gene expression regulatory systems and are central elements in determination of differentiation and development. Two distinct domains that occur within transcription factor proteins include a DNA binding domain that permits binding to sequence elements in DNA and secondly an activator domain that directly impacts transcription. In some cases the transcription factor interacts with a co-activator domain. The repertoire of transcription factors includes general or basal transcription factors that facilitate transcription and specific transcription factors that bind only to specific DNA sequences.

Transcription initiation

Control of transcription initiation is essential to development. Distal and proximal enhancers and promoters are involved in this process. Transcription start sites can be identified through analysis of sequence at the 5′ end of full-length cDNAs (complementary DNA segments sequenced from messenger RNA). Kawaji *et al.* (2006) reported that for a specific gene, transcription start site selection varied in different tissues and that most genes do not have a specific single transcription start site. They determined that for a specific gene, alternative start sites were present and these were spread across the 5′ gene region. Furthermore for a specific gene transcription, start site selection varied in different tissues.

Recent studies have revealed that transcription factors may bind at the transcription start site or downstream or upstream of that site. The transcription machinery that assembles at the promoter includes 27 polypeptides including general transcription factors.

Enhancers contain specific DNA sequence elements to which transcription factors bind. However the low sequence specificity of the 6–12 nucleotide elements that bind transcription factors results in the potential for a specific sequence element to bind different transcription factors. It is therefore possible that at different stages of development different transcription factors may bind to a specific enhancer element. Furthermore it seems likely that different combinations of transcription factors may bind to a specific enhancer site.

Pioneer transcription factors

Spitz and Furlong (2012) reviewed evidence for the existence and function of pioneer transcription factors. These bind to specific genomic sequence sites and enhancer sequences and lead to alterations in nucleosome positioning in adjacent genomic regions. The pioneer transcription factors do not necessarily activate the enhancers to which they bind. The pioneer transcription factors may subsequently be replaced by other transcription factors. Pioneer transcription factors may also serve to protect enhancer sites from methylation. Examples of transcription factors that may have pioneer function include MyoD and PAX5.

Spitz and Furlong (2012) emphasized that expression of a specific gene is dependent upon enhancers, available promoter elements and the three-dimensional arrangement of genome segments.

An important consideration is whether structural chromosome changes; e.g., duplication, deletions, translocations and inversions alter the three-dimensional chromosome organization and the potential for promoter enhancer interactions.

Transcription factors and pluripotency of cells

The field of transcription factor research took a great leap forward with the discovery of the key role of transcription factors in converting somatic cells, such as fibroblasts, to pluripotent stem cells. Stem cells and induction of pluripotency and differentiation will be discussed in a subsequent chapter.

Function and expression of human transcription factors

In a review of human transcription factors, Vaquerizas *et al.* (2009) presented data on 1,391 manually curated sequence specific transcription factors. Complete genome sequence analysis of DNA elements that bind to transcription factors have led to the creation of a number of different databases with inventories of transcription factors. Classifications of transcription factors are based on DNA binding characteristics, on protein domains and structural homologies and on the basis of the biological processes in which they participate. Vaquerizas *et al.* (2009) noted that in some cases the structural characteristics of a transcription factor provide some insight into its biological function, e.g., homeodomain transcription factors are often involved in developmental processes. Classifications are sometimes based on the tissue or tissue in which the transcription factor is expressed. It is important to note that transcription factors undergo extensive protein–protein interactions and in some cases combinations of transcription factors determine regulation.

Vaquerizas *et al*. (2009) classified transcription factors into 23 families and added a 24 "other" category for undefined transcription factors. Zinc finger transcription factors constituted the most predominant class, approximately 680 of 700, followed by homeodomain factors approximately 250 of 700 and the helix-loop-helix transcription factors 80 of 700. They also classified transcription factors according to biological function based on literature reports. Of 741 factors thus analyzed 263 were involved in developmental processes, 221 in cellular processes, 109 in metabolic processes, 66 in responses to stimuli, 30 in immune processes, 28 in reproductive processes and 24 in localization.

Vaquerizas *et al*. (2009) reported that results of expression studies revealed that approximately one third of transcription factors were expressed primarily in one tissue. These included the heart specific transcription factor NKX2-1, and the fetal brain expressed transcription factor MYCN. They reported that the central nervous system compartment, including whole brain, spinal cord and fetal brains had seven transcription factors in common. These included the thyroid hormone alpha-receptor (THRA) and the aryl-hydrocarbon receptor ARNT2.

They noted that the serum response factor (SRF) is an example of a universally expressed transcription factor that is involved in multiple processes and that frequently combines with other factors to exert its effects. Protein–protein interactions are key to co-operative interactions between transcription factors.

Chromosomal distribution of transcription factor loci

Vaquerizas *et al*. (2009) analyzed the chromosomal distribution of human transcription factor loci. They reported that 20% of these loci mapped to 23 high-density clusters. The gene loci that encode HOX transcription factors (homeodomain containing proteins) map in specific clusters and other clusters contain zinc finger transcription factors. The short and long arms of human chromosome 19 are locations of particularly striking clusters of transcription factor loci. These authors proposed that repetitive tandem duplication events led to generation of the series of paralogous, i.e., related genes. A second type of cluster occurred in the region of centromeres and telomeres. Genes in these clusters are non-paralogous. Vacquerizas *et al.* (2009) proposed that these clusters resulted from the intense genomic shuffling that occurs in these regions and genes in these clusters.

In a report on a specific transcription factor database TFClass, Wingender *et al.* (2012) provided classification of 1,558 human transcription factors and noted that if different isoforms are taken into account there are at least 2,900 different transcription factors. They noted that these factors regulate gene expression through binding to sequence elements in promoters, enhancers, silencers and other regulatory elements and emphasized that co-occurring cis-regulatory elements form comprehensive regulatory modules.

Although specific protein DNA recognition codes have been discovered for several classes, there remain classes of unclassified transcription factors. Taxonomies of DNA binding characteristics have evolved over the years and expression analysis, co-precipitation assays and bioinformatics capacities have greatly expanded. Wingender *et al.* (2012) ranked transcription factors into super-classes, classes, families and sub-families; they also included gene and peptide information. The super-class designation referred to the general topology of the DNA binding region, e.g., zinc finger transcription factors. Class designation included structural characteristics, e.g., presence of four zinc fingers. Family information included factors with structural and functional similarities.

The five most abundant sub-class categories of transcription factors defined in the classification of Wingender *et al.* (2012) include 52% with a zinc co-coordinating domain, 27% with a helix-turn-helix domain, 11% with a basic domain, 4% with an immunoglobulin type fold and 3% with other alpha-helical DNA binding domains.

Identification of transcription factor binding sites in DNA

One method for identifying DNA sites where transcription factors are bound is the CHIP Seq method. In this procedure formaldehyde is used to facilitate formation of bonds between DNA and bound protein. Following fragmentation of chromatin, antibodies against specific transcription factors are used to isolate chromatin fragments that have the specific transcription factor bound and therefore react with the corresponding antibody. Following this the antibody bound fragments are heated at 65°C or they are treated with proteinase to release the protein. The DNA fragments released may be captured on to microarrays or they may be bead captured and sequenced.

Nuclear receptors

These proteins act as transcription factors after they bind ligands such as steroid or thyroid hormones or other specific fat-soluble ligands. Ligand binding induces specific structural changes in the nuclear receptor proteins and these structural changes are necessary to activate the transcription factor activity (Mangelsdorf *et al.*, 1995).

Nuclear receptor proteins contain DNA binding domains and ligand binding domains. Ligand bound nuclear receptor proteins interact with other co-regulatory proteins. The co-regulatory proteins may be involved in chromatin remodeling and histone modifications.

Nuclear receptors play important roles during embryonic development and in later life.

Cis-regulatory elements that impact transcription

Cis-regulatory elements include enhancers, silencers, promoters and insulators.

Enhancers

Buecker and Wysocka (2012) reviewed the role of enhancers in gene regulation. They proposed that enhancers act as information integration hubs that facilitate precise spatiotemporal gene expression during embryogenesis.

Enhancers may be active or poised. The key activity of enhancers is the ability to drive gene expression at a distance. They are uncoupled from promoters and from transcription start sites. Enhancers respond to signaling cascades and can drive gene expression in multiple tissues.

Enhancers consist of long stretches of DNA, frequently hundreds of base pairs in length, and they are embedded in non-coding DNA in the genome. They can be identified in the genome on the basis of their chromatin features.

Enhancer chromatin

Enhancers contain sequences that serve as recognition elements for transcription factors. Enhancers with bound transcription factors occur in open DNA, i.e., relatively nucleosome free DNA. Nucleosome free DNA is DNAse 1 hypersensitive. However when transcription factors are bound to nucleosome free DNA, the specific DNA that is bound to protein is slightly less DNAse 1 hypersensitive. Transcription factors are bound at both active and poised enhancers.

The activity of transcription at enhancers is dependent on recruitment of co-activator proteins; e.g., EP300, histone modifiers and chromatin remodelers, e.g., chromodomain helicase binding proteins, e.g., CHD7. Active enhancer elements also bind factors that cross talk with promoters.

Buecker and Wysocka (2012) emphasized that a key difference between poised and active enhancer is the fact that active enhancers recruit RNA polymerase II and they generate short RNAs.

At poised enhancers H3K27 (histone H3lysine 27) in flanking nucleosomes is acetylated to H3K27ac, while at active enhancers it is methylated to H3K27me3.

There is growing evidence that enhancers are key determinants of tissue specific gene expression. Increase in the number of enhancer elements and their function and diversification of promoters and expansion of the protein coding repertoire expand the repertoire of cell functions and behaviors. Buecker and Wysocka (2012) noted that there are hundreds of active enhancer signatures and they are cell-type specific. Based on numbers of enhancer elements and their length it is likely that 10% of the human genome encodes enhancer elements.

There is evidence that enhancers associated with early differentiation genes are premarked. The premarked enhancers are located in regions of low nucleosome density. In these regions transcription factors and co-activators are bound to DNA elements in which H3K4me1 and H3K4me3 are enriched and H3K27 is not acetylated. When differentiation commences H3K27 acetylation occurs and RNA polymerase II associates with enhancers and produces RNA. The transition from poised to active enhancers is dependent on signaling and environmental impact. Not all developmental enhancers are premarked; some are apparently created during differentiation.

An example of activation of a poised enhancer involves activation of alpha-globin expression where eviction of the polychrome repressive complex2 by JMJD3, a lysine specific demethylase facilitates activation.

Integration of information is in part facilitated by binding of effectors of signaling pathways to DNA elements in enhancers. Cellular environment signaling effectors and epigenetic information are integrated at enhancers. There are then extensive interactions between enhancers and active promoters to constitute transcription factories.

DNA loops and regulation

Enhancer promoter interactions frequently involve the generation of DNA loops. These loops primarily involve intra-chromosomal structures. The technique chromosome conformation capture has facilitated identification of such loops. Genomic segments that are associated with each other can be captured through use of cross-linking reagents.

de Laat and Deboule (2013) emphasized that the ongoing challenge is to determine the functional relevance of interactions between proposed regulatory sequences and their target genes.

Identification of regulatory elements in the genome and relevance to disease

Analysis of the functional effects of specific regulatory segments within the genome can be achieved through deletion of these segments. Analyses of effects of deletion have revealed that deletion of a specific enhancer sequencer that is 35kb distant from the alpha-globin locus significantly reduces alpha-globin gene expression.

Genome wide association studies carried out to find DNA markers that associate with specific diseases have frequently led to identification of significant signals located in regulatory segments of the genome. These findings indicate that alterations in expression levels at specific gene loci may be the key factor in the occurrence of common diseases.

Bauer *et al.* (2013) have investigated a specific regulatory element that impacts the expression of BCL11A in erythroid cells. BCL11A acts as a transcriptional repressor of hemoglobin F (HbF) production. Specific defects in BCL11A in erythroid cells counteract HbF silencing.

Bauer *et al.* (2013) identified specific regulatory sites in intron 2 of BCL11A. These sites were 62, 58 and 55 base pairs distant from the transcription start site. These sites manifested DNAse I hypersensitivity and also histone modifications indicative of enhancer elements, they were positive for H3K4me1 and H3K27ac and negative for H3Kme3 and H3K27me3. They also determined that in erythroid cells the transcription factors GATA1 and TAL1 occupied these regulatory sites.

Bauer *et al.* (2013) then carried out extensive analysis of single nucleotide polymorphisms (SNPs) in the genomic sequence of these regulatory sites (see Figure 1.1). They determined that a variant in SNP rs1427407 showed the strongest association with HbF levels. Sequence analysis revealed that rs1427407 falls within the peak position of binding of the transcription factors GATA1 and TAL1. In studies on heterozygotes for this SNP they determined more frequent binding of the transcription factors to the G allele than the T allele. They studied developmental regulation and determined that the genome-wide association studies (GWAS) marked polymorphism markedly influenced HbF expression.

Bauer *et al.* (2013) proposed that the expression of HbF would be a therapeutic advantage. However they also stated that the cell specificity of BCL11A needed to be taken into account.

Splicing and Alternative Splicing

Multiple different protein isoforms arise through variations in the combinations of exons included in mRNA transcripts derived from a particular gene. There is evidence that in humans at least 95% of genes have multiple exons that undergo alternate splicing (Chen and Manley, 2009). The genesis of diverse isoforms that result from alternative splicing impact temporal and spatial patterns of gene expression and play key roles in cell and tissue differentiation.

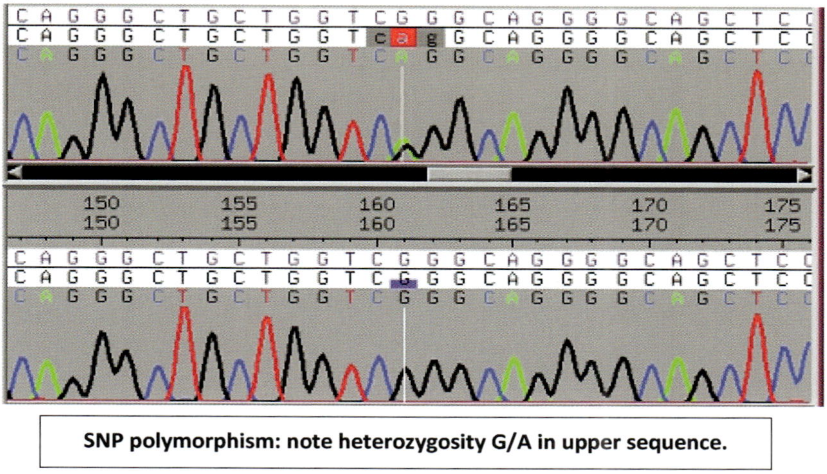

SNP polymorphism: note heterozygosity G/A in upper sequence.

Fig. 1.1: SNP DNA sequence.

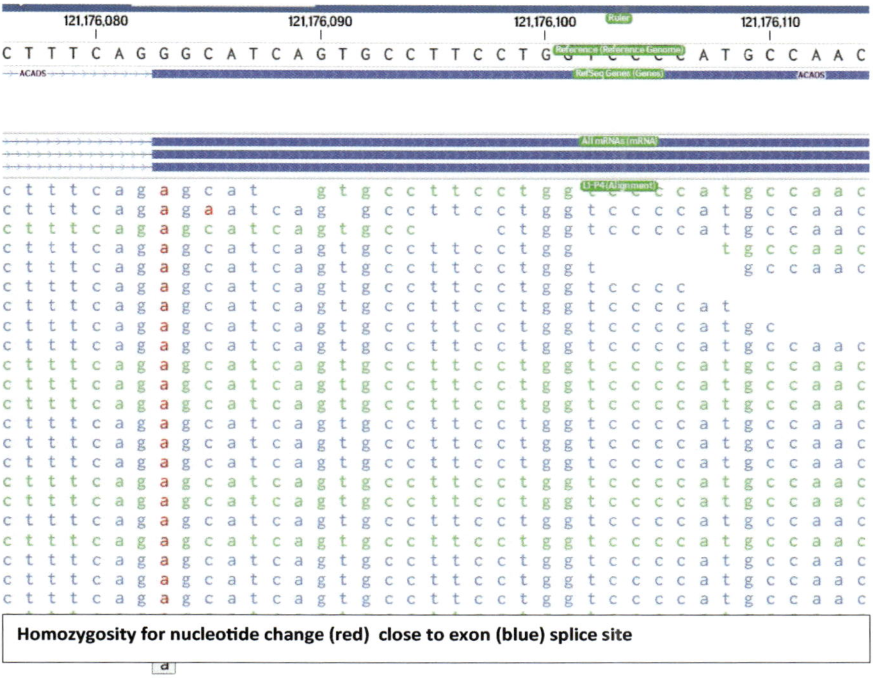

Homozygosity for nucleotide change (red) close to exon (blue) splice site

Fig. 1.2: Homozygous variant close to splice site.

Splicing is dependent on specific DNA sequences that act as recognition sequences for spliceosome binding. The key sequence at the 3′ end of the 5′ exon, the donor site is GT and the critical sequence at the start of the 3′ exon, and the splice acceptor site is AG. Also key to splicing are sequences at the branch site. This site is most frequently located within the intron. 18–40 nucleotides upstream from the 3′ splice site, the acceptor site.

An early step in splicing involves the generation of a lariat structure generated through interaction of an invariant adenosine in the branch site with the G in the splice donor site. Generation of the lariat structure is followed by nucleophilic attack, essentially a trans-esterification reaction between the OH residue in adenosine and the phosphodiester group in the splice donor site. This reaction leads to cleavage at the 5′ intron–exon boundary. Release of the 5′ exon then induces nucleophilic attack on the G nucleotide of the splice acceptor site leading to release of the lariat and ligation of the exons (Fedor, 2008).

It is important to note that cryptic splice sites also exist. These sites are generally not active or they are active at low levels and may become more active when nucleotide substitutions and abnormal functioning of the regular splice sites occur.

Splicing is dependent on the spliceosome, a complex structure composed of small nuclear ribonucleoproteins (snRNPs) and a large number of proteins involved in assembly of the complexes. At least 14 different ribonucleoproteins are included in the spliceosome complexes. One important class of protein included in the spliceosome is the serine–arginine proteins (SR proteins) and there are at least 18 different SR proteins. In addition there are proteins that play regulatory roles through binding to splice enhancers or repressors. Examples of proteins that act as splice regulators include proteins encoded by genes *NOVA1*, *NOVA*, *PTBP1* (poly-pyrimidine tract binding) and *CELF1*. Fogel *et al.* (2012) reported that the *RBFOX1* protein (RNA binding) plays a critical role in splicing during neuronal differentiation.

Abnormalities in splicing and genetic diseases

Splice site mutations are among the most frequent causes of genetically determined diseases. There is also that mutations at branch sites can lead to decrease in levels of gene products. A branch site mutation in the *NUPL*1 gene likely contributes to pathogenicity and plays a contributory role in mitochondrial complex 1 deficiency (Tucker *et al.*, 2012).

Connections between transcription and alternative splicing

Genome analysis and deep RNA sequencing have revealed that 90% of gene transcripts in humans undergo alternate splicing and that this is subject to tissue specific and developmental control (Wang *et al.*, 2008). Through inclusion of different exons in different transcripts a specific gene can produce multiple different protein isoforms that may differ with respect to their ligand binding properties, post-translational modification and cellular localization.

Luco and Misteli (2011) reviewed regulation of alternative splicing and evidence recruitment of splice enhancers, splice repressors and non-protein coding RNA to the transcriptional machinery. Their studies revealed the important of regulatory sequences located at some distance from splice sites (see Figure 1.2). They also presented evidence that the splicing machinery is recruited to the transcript prior to termination of transcription.

Processing of Transcripts at the 3′ End and Polyadenylation

Polyadenylation of mRNA transcripts is required for nuclear export, for their stability and subsequent translation. The central sequence motif for cleavage of mRNA and addition of polyA, AAUAA, was first defined in the 1970s. Additional cis-acting sequence elements have subsequently also been shown to be important.

Core elements for mRNA 3′ cleavage and polyadenylation are now known to include AAUAAA and 10 variants of this sequence, located 10–30 nucleotides upstream of the cleavage site. Downstream U and GU sequences and upstream nucleotides also impact cleavage. Polypeptide complexes important for cleavage include polyadenylation specific cleavage factors, a scaffold and polyadenylation specific polymerases. There is evidence that the cleavage stimulation factor CPSF73 acts as an endonuclease that cleavages mRNA. U1snRNP splicing factors interact with the complex. The polyA binding proteins PABPN1 and PABPN2 are also important (Elkon *et al.*, 2013).

Most human genes have been shown to have more than one polyadenylation site and there is evidence that alternative polyadenylation isoforms are abundant, Alternative polyadenylation therefore contributes to transcript diversity and adds additional layers toward regulation of gene expression.

Relevance of Alternate Polyadenylation to Differentiation and Development

Elkon *et al.* (2013) reported that transcripts in the nervous system and brain mostly have longer 3′ UTR regions than those that predominate in other tissues. Longer transcripts are derived from the use of more downstream polyadenylation sites. Proximal polyA sites are more commonly used in placenta, blood and ovary, and they give rise to transcripts with shorter 3′ UTRs. Studies on mice by Elkon *et al.* (2013) revealed that as development progresses, there is a progressive lengthening of transcript 3′ UTRs. An important finding is that when pluripotent stem cells are induced from differentiated cells there is a progressive shortening of 3′ UTRs. There is also evidence from studies of blood T cell and B cell activation that cellular proliferation is correlated with use of more proximal polyadenylation sites and generation of shorter 3′ UTRs. Together these findings indicate that alternative polyadenylation is under biological control and is of functional importance.

Derti *et al.* (2012) surveyed polyadenylation site usage in brain, liver, muscle and testes in different species. They determined that there was cross-species conservation and use of specifically positioned polyadenylation sites in a specific tissue.

Elkon *et al.* (2013) emphasized that key factors in alternative polyadenylation regulation include genes that encode proteins involved in the 3′ end processing machinery including transcription factors and the transcription elongation complex.

Aberrant polyadenylation signals are associated with specific diseases. Reduced expression of the polyA binding protein PABPN1 results in production of short transcripts due to use of proximal polyadenylation sites. Impaired PABPN1 function occurs due to triplet repeat expansion in the gene. This results in autosomal dominant oculo-pharyngeal muscular dystrophy. Specific forms of thalassemia occur due to polyadenylation site mutations. The uncontrolled proliferation of cancer cells has been shown to be associated with the use of proximal polyadenylation sites. One proposed therapeutic approach is the use of oligonucleotides that block proximal polyA sites.

Most polyadenylation sites are located in the most 3′ exon of the gene, however, polyadenylation sites sometimes occur in upstream exons and introns. Use of these polyadenylation sites potentially alters coding sequence and 3′ UTR (Tian and Manley, 2013).

Most microRNA (miRNA) target sites are located in 3′ UTR regions of mRNA transcripts. Therefore 3′ UTRs play important roles in miRNA interactions and regulation of translation.

Translation

Translation of spliced mRNA takes place on ribosomes. Ribosome formation requires adequate assembly of ribosomal RNAs and ribosomal proteins and the action of assembly factors (Karbstein, 2013). Specific assembly factors are required for small ribosomal subunit (40S) and for large ribosomal (60S) assembly.

Ribosome biogenesis

Ribosomes function to translate DNA/RNA code to proteins and their functions are critical to cell division, growth and differentiation. Factors involved in ribosome biogenesis have been discovered through studies on eukaryotes and through studies of specific human diseases. Pallade (1974) first described cellular ribosomes in the mid-50s (see Nobel Lecture reference).

Thomson *et al.* (2013) reviewed ribosome biogenesis, they noted that synthesis of ribosomes requires high cellular energy levels. Ribosomal RNA gene transcription takes place in distinct region of the nucleus, the nucleolus. In humans tandem repeat of ribosomal RNA encoding genes occur on the acrocentric chromosomes 13, 14, 15, 21 and 22. Transcription leads to generation of a precursor molecule that then undergoes modification to generate small molecules. SNO ribonuclear proteins play role in processing of the primary transcripts generated from the ribosomal RNA genes. Maturation of ribosomal RNA involves processing, folding and modification and generation of ribonuclear protein complexes. This maturation takes place partly in the nucleus but it occurs primarily after ribosomal RNA and ribonucleoproteins are transferred to the cytoplasm. Maturation involves structural changes and these require activity of ATPases and GTPases. Thomson *et al.* (2013) reported that 80 different proteins have been implicated in maturation of the 60S ribosome sub-unit.

Fully matured ribosome subunits are involved in protein synthesis. Thomson *et al.* (2013) reported that a surveillance system is in place to ensure correct ribosomal function and for defective ribosomes degradation.

Ribosome biogenesis defects

Diamond Blackfan anemia is a congenital bone marrow failure disorder characterized by macrocytic anemia and cytopenia. The disease manifests different degrees of severity and the severe forms may present soon after birth. The disorder is due to heterozygous mutations or deletions in any one of the nine different genes that encode proteins important in synthesis of the ribosomes (Teng *et al.*, 2013).

A disorder characterized by bone marrow failure and skin and nail abnormalities, and skeletal abnormalities, congenital dyskeratosis, is due to mutations in an X-linked gene *DXC1*. Teng *et al.* (2013) reported that this gene encodes a nucleolar protein associated with SNO ribonuclear proteins that function in ribosomal RNA modification and maturation.

In Treacher–Collins syndrome there are cranio-facial abnormalities (see also section on craniofacial development). This syndrome is caused by mutations in any one of three proteins involved in precursor RNA processing, POLR1D, POLR1C and TCOF1.

Mutations that impact function of a protein LAS1L which is required for processing and maturation of the 60S pre-ribosome sub-unit, lead to a congenital motor neuron disease with early lethality (Castle *et al.*, 2012). In this disease infants are at risk for early death due to respiratory insufficiency (Butterfield *et al.*, 2014). (See Table 1.1.)

Table 1.1:   Ribosomopathies.

Genes	Disease Associated
RPS7, RPS10	Diamond Blackfan anemia
RPS17, RPS19	cranio-facial, limb, heart anomalies
RPS24, RPS26	cranio-facial, limb, heart anomalies
DKC1	bone-marrow failure, skin, hair, skeletal anomalies
TCOF1	Treacher–Collins syndrome, cranio-facial anomalies
POLR1D, POLR1C	cranio-facial, laryngeal anomalies
LAS1L	motor-neuron disease

MicroRNAs

MicroRNAs (miRNAs are produced from intron of genes, from non-protein coding sequences in the genome or may sometimes be produced as sequence that is antisense to that of protein coding sequence.

The mature processed microRNAs impact translation of messenger RNA or sometimes promote its degradation. Specific miRNA sequences interact with specific sequences in messenger RNA.

MicroRNAs and regulation of expression

Variations in level of expression of miRNAs and sequence variation in miRNAs or their target genes are important contributors to the variation in levels of gene expression. MicroRNAs regulate post-transcriptional abundance of gene products. These single stranded RNA sequences are usually approximately 22 nucleotides long and bind to the target mRNA sequence. This binding may lead to degradation of mRNA or to repression of mRNA translation. There is evidence that a single miRNA may bind to multiple genes. Furthermore several different miRNAs may regulate the mRNA transcript of a specific gene (Lu and Clark, 2012). These investigators carried out analyses on polymorphic miRNA sites and identified specific single nucleotide variants (SNVs) in miRNAs that altered levels of expression of their target genes. They also noted variations in levels of expression of specific microRNAs in different individuals.

There are two sub-classes of microRNAs: canonical and non-canonical. Canonical miRNAs are 60–75 nucleotide hairpin structures that initially bind the protein encoded by the *DGCR*8 gene (encoded in the di George syndrome region on chromosome 22), and then they bind to the RNAse III DROSHA. The DGCR8 protein is essential for DROSHA activity. The DGCR8–DROSHA microprocessor complex releases the hairpin into the cytoplasm. A second RNAse III enzyme Dicer then cleaves the hairpin to 18–25 nucleotides.

Non-canonical microRNAs bypass the DROSHA complex; they are directly transcribed as short hairpins or may be transcribed as longer complexes and then cleaved by other endonucleases. Following entry into the cytoplasm, Dicer cleaves them (Babiarz *et al.*, 2008).

Proteome Analysis

Biological information accessible through mRNA analysis and proteome analyses includes information on cell type and tissue abundance. However only proteome analysis is able to provide information on post-translational protein modifications.

Wilhelm *et al.* (2014) reported that large-scale data on the human proteome were placed in a database Proteome DB. This data includes information in approximately 17,000 proteins. They defined the core proteome as between 10,000 and 12,000 ubiquitously expressed proteins that are involved in maintenance. These investigators also generated proteome profiles for 27 different human tissues and body fluids.

Kim *et al.* (2014) reported in depth, analysis of the proteome of 30 different tissues with normal histology, including seven fetal tissues. They discovered novel protein coding regions including expressed pseudogenes and upstream open reading frames. Their study revealed tissue specific proteins and these included proteins with known function and proteins, of which function was unknown. Examples included proteins with expression restricted to the brain frontal cortex that included known functional proteins, e.g., SYNGAP 1 (synaptic RasGTPase) ICAM5 (intercellular adhesion protein), SCN1A (sodium channel), SHC3 (Src homology domain signaling from neurotrophin) and CACNG3 (calcium ion channel neuronal G3). Proteins with unknown function that are present in the cerebral cortex include C8ORF46, KIAA1211L.

In their studies of fetal tissue, they identified 735 genes for which the relative rates of expression were 10-fold higher than in adult tissues. In addition they identified 40 different proteins where expression was restricted to in fetal tissue or placenta. These included well-known oncofetal antigens such as alpha-fetoprotein and insulin-like growth factor 2 binding protein (IGF2BP3) and embryonic and fetal hemoglobins (see Table 1.2).

Table 1.2: Gene products that are found only in fetal tissue or placenta.

Symbol	Gene and Protein Name/Function
HBG2	Hemoglobin gamma 2
HBG1	Hemoglobin gamma 1
HBZ	Hemoglobin zeta (yolk-sac)
AFP	Alpha-fetoprotein
MYL4	Myosin light chain 4 (embryonic atrium)
CYP19A1	Steroid metabolism, growth stimulation
SULT1E1	Sulfotransferase (estrogen preferring)
PAEP	Progesterone preferring (pregnancy associated endometrium)
HMGCR	3-hydroxymethylglutarylCoA reductase
SI	sucrose isomaltase
IGF2BP3	Insulin like growth factor 2 binding protein3 (nucleolus)
RBP2	Retinol binding protein
MYL7	Myosin light chain regulatory
MT1H	Metallothionine 1H
DCX	Doublecortin X encoded (neuronal migration)
HSD17B1	17-Beta hydroxysteroid dehydrogenase
HBM	Hemoglobin M (thought to be pseudogene but fetal expressed)
AMH	Anti-Mullerian hormone (mediates male sexual differentiation)
ENDOU	Endonuclease poly-U active

(*Continued*)

Table 1.2:   (*Continued*)

Symbol	Gene and Protein Name/Function
HBE1	Hemoglobin epsilon yolk-sac)
HEMGN	Hemogen (erythroid differentiation)
CAPN6	Calpain 6 (cysteine protease highly expressed in placenta)
HMGA2	High mobility AT hook transcriptional regulator
MAMDC4	MAM domain containing endosomal glycoprotein
DUSP9	Dual specificity protease 9 (negatively regulates Map kinase)
TFPI2	Tissue factor pathway inhibitor (serine protease inhibitor)
TRIM55	Tripartite domain ontaining (regulatory in sarcomere assembly)
PAGE4	P antigen (placenta)
LPPR3	Lipid phosphatase (brain, neurite migration and retraction)
MATN1	Matrilin (cartilage, extra-cellular matrix)
CYP2W1	Cytochrome p450 mono-oxygenase (fetal gut)
ISLR2	Immunoglobulin superfamily leucine rich (transmembrane)
UHRF1	Ubiquitin-like with ring fingers (DNA binding Hub protein)
NPPA	Natriuretic proteinA (electrolyte homeostasis fetal heart)
TYMS	Thymidylate synthetase (cell growth)
PSG4	Pregnancy specific glycoprotein 4 (placenta)
MT1E	Metallothionine1E
PHLDA2	Pleckstrin homolog like (imprinted in placenta)
CKS2	Protein kinase regulatory subunit
PPP1R1A	Protein phosphatase inhibitor

Kim *et al.* (2014) also analyzed protein isoforms. They reported that one-third of annotated genes give rise to multiple isoforms. They identified isoforms specific peptides derived from 2,450 proteins. An example is the FYN protein kinase gene that gives rise to isoform 1 in brain, and isoform 2 in hematopoietic cells, and the two isoforms have distinct functional properties.

They identified upstream open reading frames that gave rise to peptides and peptides derived from regions that have previously been described as 5′ untranslated regions. For example a region described as the 5′ untranslated region in the SLC35A4 gene produced peptides. Also alternate reading frames and 3′ untranslated regions were found to produce peptides. Kim *et al.* (2014) also identified proteins derived from transcripts designated as non-coding RNAs. They determined that 140 genes defined previously as pseudogenes produced proteins. The pseudogene *VDAC1P7* (voltage dependent ion channel) was translated to proteins in 30 different tissues, The MAGEB6P1 (melanoma antigen) pseudogene produced a peptide only in testis.

Analyses of the N-termini of proteins enable a more accurate determination of the transcription start sites.

CHAPTER 2

EPIGENETICS

Definition and Mechanisms

Induction of differentiation of diverse cell and tissue types involves differential gene expression. Epigenetics includes the analysis of regulatory mechanisms that impact genes expression that do not involve changes in nucleotide sequence but are dependent on modifications in DNA methylation, and modifications of histones and chromatin structure.

DNA methylation

DNA methylation was originally thought to be indicative of suppression of gene expression and much emphasis was placed on the 5′ methylation of Cytosine at Cytosine–Guanine residues (CpG), particularly in CpG rich regions of the genome (CpG islands) when these occurred. More recent evidence suggests that the relationship between methylation and transcription is more complicated. In summarizing key points regarding DNA methylation, Jones (2012) noted that most CpG rich regions near transcription start sites are usually not methylated and that CpG island methylation is mainly associated with long term silencing of gene expression such as the one which occurs on the inactive X chromosome, in imprinted gene region and in regions in the genome rich in transposons (DNA sequence elements that can readily move in the genome).

Jones noted further that DNA methylation occurs not only at promoter regions but also in internal regions (gene bodies). Furthermore there is evidence that methylation and demethylation are dynamic processes. Methylation of DNA is carried out through the activity of DNA methyltransferases DNMT1, DNMT3A and DNMT3B (cytosine methyltransferases). DNMT1 is required for maintenance methylation.

DNA demethylation can be achieved through active or passive mechanisms. Active demethylation requires activity of TET enzymes. These enzymes catalyze the hydroxylation of 5-methyl cytosine to yield

5-hydroxymethyl cytosine that then undergoes base excision repair to yield unmodified cytosine (Kohli and Zhang, 2013). Passive demethylation includes down-regulation of DNA methylases.

Examples of dynamic changes in DNA methylation in brain

Lister *et al.* (2013) analyzed the dynamics of DNA methylation in the brain frontal cortex at different stages of development. Evidence for the importance of cytosine methylation in brain development was obtained in prior studies that involved deletion of DNA methyltransferase genes *Dnmt1* and *Dnmt2* from brain genes in gene specific knockouts in mice. Lister *et al.* (2013) examined methylation of cytosine-guanine nucleotides (CG methylation). In addition they studied cytosine methylation outside of CG nucleotides. This methylation involves primarily cytosine adenine (CA) nucleotides.

Their studies revealed that accumulation of non-CG methylation occurred in neurons during the developmental phases of synaptogenesis. Cell-type specificity of methylation in intragenic regions was also observed.

They noted that methylation outside of CG residues is much more common in brain than in other tissues. Specific areas of increases in accumulation of non-CG methylation include the middle frontal gyrus. Genome regions that do not include non-CG methylation occurred in genes that involved sensory neuron function and genes involved in immune response. Levels of non-CG methylation were higher in glia, while in neurons 53% of methylation occurred in non-CG regions and 47% as CG related methylation. Cell-type specific differences in levels of CG methylation *versus* non-CG methylation were observed.

Histones and Methylation

Five different forms of histone occur in human chromatin. Histone H2A, H2B, H3 and H4 form the core proteins of nucleosomes and H1 and H5 histone form the links between nucleosomes. Histones H3 and H4 have long tails and these tails can undergo a series of different modifications including methylation, phosphorylation, ubiquitination and sumoylation.

Methylation occurs primarily at lysine residues in histones, however it can also occur at arginine residues and rarely at histidine residues. Lysine residues (K) can be monomethylated, dimethylated or trimethylated. Arginine residues can be monomethylated or dimethylated.

The most important histone modifications that impact expression include H3K27 (histone 3 lysine27), H3K4, H3K9, H3K36 and H4K20. A large number of different proteins with histone methyltransferase activity have been identified. Different forms of methyltransferase methylate different residues and one, two or three methyl groups may be added to a specific amino acid residue. The protein arginine methyltransferases (PRMT1–5) are distinct from those that methylate lysine. Methylation residues are transferred from S-adenosylmethionine to histones.

Different families of histone demethylases exist. Specific demethylase may be required to remove methyl groups from specific sites, e.g., KDM6A and KDM6B remove methyl groups from H3K27. However KDM4A and KDM4B may remove methyl group from different lysine sites, e.g., H3K9 and H3K36.

Specific proteins facilitate recruitment of histone methyltransferase or histone demethylases to specific sites. These proteins include polycomb proteins and non-coding RNAs may also be involved. In addition there are specific proteins that recognize methylated histones. These include chromodomain proteins, bromodomain proteins, Tudor domain and PHD finger proteins. These are sometimes referred to as chromatin readers and will be discussed further in the section on chromatin modification.

H3K4me3 is generally associated with active transcription and H3K27me3 is often associated with repression of expression. However, Greer and Shi (2012) noted that histone modification alone is often insufficient to modify activity and the binding of additional proteins is important. Silencing of gene expression also occurs as a result of methylation of lysine 27 in histone H3 (H3K27).

Enzymes and Proteins Important in Chromatin Modification

Kouzarides (2007) noted that the timing of modification is dependent upon signaling conditions in the cells. Specific enzymes have been identified for methylation, demethylation, acetylation, phosphorylation, ubiquitination and sumoylation, ADP ribosylation, deimination and proline isomerization.

Kouzarides (2007) reported that active chromatin and silent chromatin are associated with different sets of modification and are often separated by boundary elements including CTCF transcription factors. Active chromatin has high levels of acetylation and high levels of trimethylated H3K4, H3K36 and H3K79.

Deimination leads arginine to be converted to citrulline and this conversion inhibits its activation. Reversal of acetylation occurs during transcriptional repression. Lysine methylation at sites H3K9, H3K27 and H4K20 are associated with transcriptional repression. Proline isomerization involves conversions between cis and trans forms. This process leads to structural alteration in proteins, including histones, and inhibits methylation.

Ubiquitination involves primarily histone H2 and Kouzarides (2007) noted that the exact function of this was not known.

Chromatin Architecture and Gene Expression

The position and density of nucleosomes in a specific genome region and the presence of active or repressive histone variants constitute the chromatin architecture. Nucleosomes can be remodeled through exchange of specific histone variants and histones through chemical modifications.

Proteins involved in modification and function of nucleosomes are sometimes classified as writers, erasers and readers (Helin and Dhanak, 2013). Writers modify histones through methylation, acetylation, phosphorylation and ubiquitination, and include histone acetylases, histone kinases and ubiquitin ligases. Readers interpret DNA and histone modifications. Readers are contained within protein complexes including proteins with bromo, chromo and Tudor domains. Erasers remove post-translation modification and they include histone deacetylases, histone demethylases and phosphatases.

Chromatin remodeling

Chromatin remodeling involves the use of ATP hydrolysis to provide energy to alter nucleosome position. BRG1 associated factor (BAF) complexes

Table 2.1: Protein Subunits and Encoding Genes.

Protein Subunit	Encoding Gene
BAF250a	*ARID1A*
BAF250b	*ARID1B*
BAF47	*SMARCB1*
BRM	*SMARCA2*
BRG1	*SMARCA4*
BAF155	*SMARCC1*
BAF170	*SMARCC2*
BAF180	*PBRM*
BAF45A	*PHF10*
BAF53A	*ACTL6A*
BAF53B	*ACTL6B*

play important roles in chromatin remodeling. These complexes include 15 subunits and core ATPase subunits (see Table 2.1).

Ronan *et al.* (2013) reviewed the roles of BAF complexes in development of the nervous system. BAF53B is required for dendritic morphogenesis. Neuronal circuitry development involves BAF53A. In vertebrates, BAF complexes correspond to SWItch/Sucrose Non-Fermentable (SWI/SNF) complexes in yeast.

Koga *et al.* (2009) reported that specific *SMARCA2* risk alleles were reported to be associated with schizophrenia and *SMARCA2* encodes BRM. BRM protein forms a network with other genes reported to be associated with schizophrenia.

Other ATP dependent chromatin remodeling proteins

CHD8 (chromatin helicase domain 8) is an ATP dependent chromatin remodeling protein found mutated in 13 cases of autism spectrum disorder in large scale sequencing experiments. The mutations were frameshift, indels (insertions or deletions) or nonsense mutations and macrocephaly frequently occurring in the affected children. In reviewing these findings, Ronan *et al.* (2013) noted that brain overgrowth is likely due to the fact that decreased CHD8 activity leads to WNT beta-catenin signaling pathway upregulation with increased proliferation.

Chromatin modifiers

Ronan *et al.* (2013) noted that mutations in a number of other chromatin modifiers have been reported to be associated with neurodevelopmental disorders. Specific examples include mutations in the CREBBP (cyclic AMP response element binding protein) that has histone acetyltransferase activity and mutation in EP300 that binds to CREBBP in Rubinstein–Taybi syndrome, associated with congenital malformation and intellectual disability. Histone methyltransferase functional defects due to mutations in KMT2A (MLL1) occur in Wiedemann–Steiner syndrome, and in Kleefstra syndrome, EHMT1 lead to syndromes with intellectual impairment. EHMT1 (euchromatic histone lysine N-methyltransferase) plays a role in trimethylation of H3K9. In Kabuki syndrome, that may also be associated with intellectual impairment, mutations occur in histone methyltransferase KMT2D. In some cases of Kabuki syndrome there are mutations in the histone demethylase KDM6A. Histone lysine demethylase defects, due to mutations in PHF8 (PHD finger protein 8) that demethylates H3K9, lead to X linked mental retardation. Defects in histone ubiquitination occur in case of structural changes or mutations in the X linked gene *HUWE1* that encodes an ubiquitin protein ligase.

Additional evidence of cognitive impairments associated with defective BAF subunit function

Coffin–Siris syndrome is associated with cognitive impairment, microcephaly, coarse facial features and hypoplastic nails on 5th finger or toe. In patients with this syndrome mutations have been found in genes that encode BAF subunits. Tsurusaki *et al.* (2014) carried out exome sequencing in 23 individuals with this syndrome. In individual patients they identified germline mutations in one of six BAF subunits encoding genes *SMARCB1, SMARCA4, SMARCA2, SMARCE1, ARID1A* and *ARID1B*. In one patient an interstitial deletion in *SMARCA2* was found on single nucleotide polymorphism (SNP) microarray. Haploinsufficiency of these gene products apparently leads to developmental abnormalities.

Defect in the genes that encode BAF complex subunits occur in the Nicolaides–Baraitser syndrome, characterized by short stature, unusual facial features, speech impairments and autism. Sousa *et al.* (2009) described defects in the *SMARCA2* gene that encodes the BRM subunit in patients with this disorder.

Mutations in genes encoding BAF subunits have been found in patients with autism. These include mutations in *SMARCC1* (BAF155), *SMARCC2* (BAF170), *PBRM* (BAF180) and *ARID1B* (BAF250b).

Dendritic spine abnormalities are key findings in autism and schizophrenia. Studies have revealed that the BAF53b subunit encoded by *ACTL6B* play key roles in neuronal branching and synapse formation, and in determining synaptic plasticity, and learning and memory.

Different forms of histone deacetylases differ in their impact on learning and memory. In mice HDAC4 deletions leads to brachycephaly and mental retardation. There is evidence that deletion of HDAC4 is a key factor leading to brachycephaly and mental retardation in patients with the 2q37.3 deletion (Williams *et al.*, 2010).

Ronan *et al* (2013) raised the question why mutations in chromatin regulation particularly impact the nervous system given that chromatin regulation functions widely. They proposed that fine-tuning of chromatin regulation and gene expression may be particularly important in the central nervous system.

Polycomb repressor complexes

Polycomb repressor complexes PRC1 and PRC2 participate in post-translational modification of histones. These complexes play key roles during development and impact cell differentiation and cell fate decisions (Morey and Helin, 2010). Loss of PRC2 in mice leads to embryonic lethality. PRC1 deficiency does not lead to embryonic lethality but does lead to developmental defects.

PRC2 participates in the di- and tri -methylation of lysine 27 in histone H3 (H3K27me2/3). The catalytic subunit of PRC2 is enhancer of zeste *EZH2*. *EZH2* mutations or deletions were identified in Weaver syndrome, a disorder characterized by dysmorphology, learning disabilities and overgrowth including macrocephaly. This overgrowth may be related to imbalance in the WNT beta-catenin system (Ronan *et al.*, 2013). *EZH2* interacts with two other subunits in PRC2, namely SUZ and ECD, Other subunits present in PCD2, not required for catalytic activity include HDAC1 and HDAC2.

PRC1 participates in mono-ubiquitylation of lysine 119 in histone H2A. The catalytic subunits of PRC1 are RING1A and RING1B. The activities of PRC1 and 2 include generation of H3K27me and H2AK119Ub1, and lead to chromatin compaction and inhibition of transcription elongation. Morey and Helin (2010) reported that PRC1 and 2 regulate expression of thousands of genes. Key to this regulation is binding of PRC to specific Polycomb responsive DNA elements. These elements occur in gene promoters and are sometimes located many kilobases distant from transcription start sites.

Specific non-protein coding long RNAs apparently recruit Polycomb complexes. These include HOTAIR, a long RNA that suppresses *HOX* gene expression. The XIST RNA that suppresses expression of genes on one X chromosome in females recruits PRC2. Co-repressor proteins associated with the polycomb proteins include BCOR. The BCOR protein is mutated in X-linked oculofacio–cardio–dental syndrome (OFCD). Morey and Helin (2010) noted that the protein JARID2 (Jumonji C and ARID domain containing) interacts with PRC2 and facilitates targeting of this complex to specific genes.

The presence of specific histone variants such as histone H2AZ and macro H also modify chromatin function.

Readers

Reader proteins carry out the functional interpretation of post-translational histone modifications. The binding of chromatin readers to DNA impacts DNA transcription, replication and DNA repair (Musselman *et al.*, 2012). Different types of reader proteins are known, each type differs with respect to the type of modified histone to which it binds (Lu *et al.*, 2013). Bromodomain containing readers bind to histone with acetylated lysine residues. Reader proteins 14-3-3 and BRCT recognize histones in which serine or threonine is phosphorylated. The plant homeodomain (PHD) proteins contain zinc fingers that recognize histone H3 methylated on lysine 4.

Table 2.2:   Reader Proteins and Histones.

Readers	Histones
Methyl-lysine readers	H3K9me3/2, H3K27me2, H3K4me3, H3K9me3
Chromodomain, TTD	H3K3me2, H3K20me2
Methyl-arginine readers	H3Rme2, H4Rme2, H3R2me2
Tudor	H3K4me3
Acetyl-lysine readers	H3Kac, H4Kac, H2Akac
Bromodomain	Acetylated histones
Phosphoserine readers	H3S10ph
Phosphothreonine readers	H3T3ph
14-3-3, BRCT	H2 phosphohistone
Unmodified histone readers	H3un
PHD finger proteins	H3K4me3 binding

Another family of proteins, the Tudor and the chromodomain proteins interact with methylated groups present in tail histones (see Table 2.2).

Non-Protein Coding RNA

There is evidence that non-protein coding RNAs impact gene expression. These include: Long non-coding RNA (Linc or lnc RNA), Anti-sense transcripts, Short non-coding RNAs (200 nucleotides or less in length), Endogenous short inhibitory RNAs (siRNAs), SnoRNAs and MicroRNAs.

Lnc RNAs are transcribed from intergenic and from intragenic regions. They range in size from 200 nucleotides to 100kb.

Natural antisense transcripts are transcribed starting from the opposite end of the protein coding transcript and their sequence is complementary to that of the sense strand. Natural antisense transcripts play roles in gene regulation and they provide a scaffold for interactive proteins (Magistri *et al.*, 2012; Mattick, 2012).

XIST was the first antisense non-coding transcript identified. There is evidence that XIST recruits the polycomb repressive complex PRC2 and that this leads the XIST coated X chromosome to undergo heterochromatization and repression of expression.

Other examples of antisense non-coding transcripts include transcripts derived from sequences in imprinted genomic regions. One example is KCNQ1OT, an antisense transcript transcribed from sequence in intron 10 of the *KCNQ1* gene on the paternally derived copy of chromosome 11. This antisense transcript KCNQ1OT1 acts in cis to silence KCNQ1 and neighboring genes. The silencing is mediated through KCNQ1OT1 promotion of histone methylation and through recruitment of polycomb repressive complexes.

Expression of non-coding RNA is high in the brain. Magistri *et al.* (2012) noted that levels of brain derived neurotrophic factor (BDNF) are controlled by antisense transcripts (BDNFAS) that alter chromatin modification in the *BDNF* gene. Modarresi *et al.* (2012) also described antisense transcripts that impact levels of glia derived neurotrophic factor (GDNF) and of Ephrin receptor 2.

Natural antisense transcripts usually exert their regulatory effects in cis, however there are examples of natural antisense transcripts that have trans effects. The lnc RNA, HOTAIR is encoded from sequence on chromosome 12 and it is antisense to the HOXD loci. HOTAIR associates with sequences in the HOXD locus on chromosome 2 and silences expression of HOXD through binding of the PRC2 and of co-repressor complexes and through increased methylation.

There is evidence that lnc RNAs play roles in the maintenance of pluripotent states of stem cells. Specific repeat elements in lnc RNA sequences are apparently particularly important in their binding to DNA sequence and induction of repression. One example of such a repeat element is the REPA element that occurs in the XIST sequence.

Genomic Imprinting

Genomic imprinting is an epigenetic process that leads to monoallelic parent specific expression of genes that map to specific chromosome regions. Imprinting was discovered in mice that revealed that normal embryogenesis required both male and female genomes (McGrath and Solter 1984). Discovery of imprinting grew also in part from studies that revealed that uniparentaldisomy in specific chromosome regions led to congenital malformations (Yamazawa *et al.*, 2010). In uniparentaldisomy, an individual inherits both copies of a specific chromosome or a specific chromosome segment from one parent and no copies from the other parent.

Thus far, more than 100 loci are imprinted in mice and many of the corresponding loci in humans are also imprinted.

Arnaud (2010) reviewed genomic imprinting. Most imprinted genes lie within large clusters of genes. Each cluster contains genes that are expressed exclusively from maternal derived alleles, genes expressed exclusively from paternal derived alleles, non-imprinted genes expressed from both parental alleles and genomic sequences that express non-protein coding RNAs. Expression of the genes that are present in these clusters is often regulated by cis-acting elements at imprint control regions (ICRs). Epigenetic modifications occur at ICRs.

ICRs and epigenetics

ICRs may extend over several kilobases and within these regions, there is a differentially methylated region (DMR). The DMR carries methylation passed on either from the male germ cell or from the female germ cell. Arnaud (2010) emphasized that the DNA methylation imprint passed on from the germline is maintained in somatic lineages throughout development. There is evidence that activity of a DNMT3A/DNMT3L complex is required for methylation and imprinting in the germline.

In early embryogenesis, global demethylation of DNA occurs. However, the parental specific germline imprints must be maintained. Memory of the parental origin of specific imprinted chromosome regions must be maintained in somatic cells.

Methylation of the DMR region leads to silencing, however, for long ICRs, non-coding RNAs and antisense RNAs are required. ICRs exert their effects through DNA methylation, histone modification, and expression of non-coding RNA, alteration of insulator binding and alteration of higher order chromatin structures.

In humans, analysis of specific chromosome deletions led to mapping of ICRs. Deletion of an imprint control locus from the specific chromosome on which it is normally expressed, leads to dysregulation of the associated gene cluster.

Imprinted genes are often involved in growth, neurodevelopment and behavior. Abnormalities in these processes occur when imprinted gene regions are disrupted by deletion, duplication, inversion, translocation or point mutations.

Imprinting disorders due to defects in the 15q12–q13 region

Angelman syndrome (AS) is due to deficiency of maternally expressed UBE3A, particularly in the central nervous system. In many cases, it is due to chromosome deletion but it may also arise through mutations in the *UBE3A* gene; in some cases it is due to mutation in the AS ICR.

Prader–Willi syndrome (PWS) arises most frequently as a result of deletions that remove the paternally expressed genes in the 15q12–q13 regions. In 1–3% of cases, this syndrome arises as a result of abnormal imprinting in the PWS imprint control locus in the *SNRPN* gene cluster.

Transient neonatal diabetes

There is a form of diabetes mellitus that presents in the newborn period and that may remit after a few months but may recur later in life. Genetic studies revealed that this disorder is frequently associated with paternal uniparentaldisomy of a region on 6q24; in some cases duplication of 6q24 on the paternally derived chromosome was found (Docherty *et al.*, 2013). Over-expression of the 6q24 located genes *PLAGL1* and *HYMA1* occurred. Hypomethylation of genes in this region and at multiple loci was found in some cases.

Pseudohypoparathyroidism

The ICR of the *GNAS* gene (Guanine nucleotide neuroendocrine secretory protein) and its cis control element are usually hypermethylated on the maternal chromosome (Turan and Bastepe, 2013). Loss of methylation in the *GNAS* control region leads to pseudo-hypoparathyroidism and renal resistance to parathyroid hormone.

Hydatidiform mole

Complete hydatidiform mole is characterized by hyper-proliferation of the trophoblast and the absence of fetal development. It usually arises as a result of fertilization of an egg from which the nucleus is absent; the paternal genome undergoes diploidization. This process is referred to a diploid androgenesis (Fisher *et al.*, 2000).

Biparentalhydatidiform moles sometimes referred to as partial moles have trophoblastic hyperplasia and partial abnormal fetal development. These moles may arise in cases of dispermy triploidy.

Recurrent hydatidiform moles or inherited predisposition to hydatidiform moles occur as autosomal recessive conditions due to mutations in the *NLRP7* gene (NLR family pyrin domain containing). Cases are homozygous or compound heterozygous for deleterious mutations in *NLRP7* gene. Dixon *et al.* (2012) reported that non-synonymous deleterious *NLRP7* gene mutations have also been found in women with a poor reproductive history and pregnancy losses. The *NLRP7* gene is apparently required to establish methylation imprints at multiple loci. There is evidence that *NLRP7* function is particularly important for chromatin function in early embryogenesis (Mahadevan *et al.*, 2014).

Beygo *et al.* (2013) studied DNA methylation in a patient with multiple congenital malformations born to a mother who was heterozygous for a *NLRP7* gene mutation. The patient was homozygous for alleles in the NLRP7 region and had methylation defects at imprinted loci.

Diseases due to Defects in lnc RNAs

Lnc RNAs may be derived from large intergenic regions that may give rise to long transcripts, e.g., 50kb in length and to smaller transcripts. Functions of lnc RNAs include participation in regulation of expression due to the fact that they bind to DNA and act as a scaffold for the binding of regulatory proteins. A number of lnc RNAs are associated with epigenetic regulation.

Examples of disease-associated lnc RNAs include diseases that involve TERC. TERC is the RNA component of telomerase and is encoded on chromosome 3. The protein component of telomerase is encoded on chromosome 5. Defects in TERC lead to dyskeratosis congenital (Lu *et al.*, 2012).

There is evidence that a lnc RNA plays a role in the pathogenesis of Facio-scapulo-humeral muscular dystrophy (FSHD). This disorder is associated with reduction in the length of repetitive element D4Z4 that maps to chromosome 4q35. Pathogenic contraction of D4Z4 is associated with inappropriate increased expression of *DUX4* gene that maps within an associated repeat array. *DUX4* gene is normally expressed only in testis and it is epigenetically repressed in somatic tissues. In FSHD the loss of D4Z4 repeats leads to decreased repression of *DUX4* and to *DUX4* expression in skeletal muscles. In addition DUX4 protein activates expression of other genes and retrotransposons (Young *et al.*, 2013).

Interaction of fetal and maternal intergenic genotypes:
HELLP syndrome

HELLP syndrome occurs during pregnancy and is associated with hemolysis, elevated liver enzymes, low platelet count and placental dysfunction. There is evidence that the syndrome is dependent upon the interaction of specific fetal and maternal genotypes. The disease maps to a small intergenic region on chromosome 12q23.2 that encodes an lnc RNA that is expressed by trophoblasts and regulates multiple genes that are related to trophoblast proliferation. Loss or mutation of this lnc RNA leads to increased trophoblast proliferation and invasion. There is evidence that homozygosity for defects in the HELPP locus leads to the syndrome. Homozygosity results from inheritance of the risk locus from the mother and defects in the paternally derived gene copy in the fetus; these defects may be *de novo* or inherited (Van Dijk *et al.*, 2012).

Separating genes from their upstream control regions:
Brachydactyly

Mutations in the chromosome 12q23.2 gene *PTHLH,* a member of the family of parathyroid hormones, and mutations in the *SOX9* gene on chromosome 17 can lead to brachydactyly. Maass *et al.* (2012) identified a regulatory region at 24Mb distant from *PTHLH* that controls regulation of that gene and that includes sequences for lnc RNA. They determined that a regulatory element CISTR and the lnc RNA DA125942 upstream of *PTHLH* are controlled by expression of a product derived from the *SOX9* gene. Analysis of patients with translocations that impacted 12q23.2 revealed that when *PTHLH* and its upstream regulatory element CISTR and lnc RNA were separated, expression of *PTHLH* was not regulated and this lead to premature chondrocyte differentiation at the expense of bone elongation, leading to brachydactyly.

ATRX deficiency

ATRX deficiency leads to cognitive impairment, facial, skeletal and uro-genital abnormalities and mild thalassemia. Alpha thalassemia arises due to down regulation of alpha-globin gene expression. ATRX protein normally localizes to tandem repeat sequences upstream of the alpha-globin genes and apparently facilitates expression of alpha-globin genes. There is evidence that ATRX protein has wide-spread effects. It interacts with DAXX (death domain protein) that has multiple functions in apoptosis and transcription repression. ATRX protein also functions as a chaperone complex to deposit histone variant H3.3 into pericentric, telomeric and ribosomal repeat sequences (Clynes *et al.*, 2013).

Transgenerational epigenetic effects

Specific genetic or environmental factors may impact the uterine environment in a specific pregnancy. Waterland and Michels (2007) proposed that reduced availability of methyl groups, through nutritional or environmental effects, might alter DNA methylation. This might lead to epigenetic modifications of germ line cells in the fetus. The epigenetic effects could potentially manifest in the offspring of the fetus.

Epigenetic silencing mechanisms are important in silencing repetitive elements. Failure to silence these elements may lead to genomic instability.

Epigenetic Factors in Generating Pluripotent Stem Cells

Takahashi and Yamanaka (2006) reported that pluripotent stem cells were generated from somatic cells through the actions of the transcription factors Oct4, Sox2, Kif2 and c-Myc (OKSM).

There is evidence that the transformation of somatic cells to pluripotent stem cells (iPSCs) erases somatic epigenetic signatures including DNA and histone modification at specific loci. In addition new epigenetic marks appear.

Doege *et al.* (2012) reported that two epigenetic factors are essential for pluripotency induction these are the enzymes PARP1 and TET2. PARP1 regulates modification of 5-methylcytosine. TET2 functions in the oxidation of 5-methylcytosine and generation of 5-hydroxymethylcytosine. These investigators reported that an early stage in induction of pluripotency involved the recruitment of PARP1 and TET2 to the *NANOG* and *ESR1B* (estrogen receptor related) (Esr1b) loci. This recruitment led to enrichment at these loci of the activation associated histone modification H3K4me2 and decrease in the silencing marker H3K27me3.

Gifford *et al.* (2013) compiled extensive maps of chromatin modifications and DNA methylation in pluripotency and in subsequent generations of differentiated cells.

Epigenetics, pluripotency differentiation and non-protein coding elements

Bernstein *et al.* (2006) reported that highly conserved non-protein coding elements (HCNEs) cluster in genomic regions that encode transcription factors are important in development.

They also emphasized that modifications of histone proteins play important roles in epigenetic regulation. Histone H3 lysine 4 (H3K4) and Histone H3 lysine 27 (H3K27) methylation modifications are particularly important in regulating gene expression. H3K4 recruits nucleosome remodeling enzymes and histone acetylases and increases transcription. H3K27 methylation enhances compaction of chromatin and negatively impacts transcription.

Analysis of embryonic stem cells has provided important insights into factors that promote pluripotency and changes that emerge with differentiation. Bernstein *et al.* (2006) reported that approximately 200 HNCE regions occur in mammalian genomes. These regions tend to be gene poor but are enriched for genes that encode developmentally significant transcription factors. Clusters of *HOX* genes also occur in these HCNE regions.

Bernstein *et al.* (2006) mapped histone methylation patterns in the HCNEs. They identified bivalent domains within HCNEs. These consisted of large regions of H3K27 methylation and small regions of H3K4 methylation. Based on their studies of bivalent domains and gene expression, Bernstein *et al.* (2006) proposed that bivalent domains silence developmental genes in embryonic stem cells while keeping them poised for activation.

They analyzed histone methylation status of the 332 known transcription factors. Differentiation of stem cells into differentiated cells involves increased H3K4 levels. Differentiation occurs in response to environmental cues and specific gene expression patterns.

CHAPTER 3

SIGNALING SYSTEMS

Ligand Receptor Interactions and Downstream Pathways

In signaling pathways an extracellular ligand interacts with a receptor that is most often a transmembrane protein structure. Ligand binding activates receptor-associated domains that are within the cell and triggers a series of downstream reactions that frequently result in expression of specific genes.

The activated receptor domains may produce signal directly. In most cases the activated receptor then activates a series of proteins that give rise to second messengers.

Some receptors have kinase domains in their cytosolic segments. On binding of ligand to receptor autophosphorylation occurs and then receptor can activate a target protein by phosphorylation.

Some receptors are associated on their cytosolic side with guanine nucleotide binding proteins (G proteins). Inactive receptors are associated with the inactive form of the G protein that is bound to a nucleotide diphosphate (GDP). On receptor activation the bound nucleotide is in the triphosphate form (GTP). The activated form of the G protein then reacts with other target proteins in the cytoplasm (Lewin, 2000).

Activated second messengers may subsequently activate a series of proteins in a signaling cascade. The end reaction of the cascade is generation of a molecule that enters the cell nucleus and alters the activity of a transcription factor that impacts gene expression.

There are different G proteins with different subunit structures most commonly alpha, beta and gamma subunits and the subunits differ in their sequence. Different G protein subtypes interact with different second messengers. Different reactions result from activity of different G proteins. The reactions include interaction with cyclic AMP, with specific lipids, stimulation of inositol triphosphate or diacylglycerol components and release of calcium.

Much progress has been made over the past several decades in identifying and analyzing cell signaling systems. Key remaining questions relate to regulation and spatio-temporal co-ordination of expression of signaling. Scott and Pawson (2009) delineated important mechanisms that facilitate co-ordination. These included the existence of multi-protein complexes. They noted that in some cases signals were processed through multi-protein complexes, in other cases multi-protein complexes were formed in response to signaling. Key factors in the development of multi-protein complexes included post-translational modification of proteins and particularly phosphorylation. They noted further that the structural features of signaling proteins facilitates modification, and that interacting, binding and docking domains are distinct from catalytic domains.

They examined domains that are key elements in protein interactions and reported that Src homology domains, e.g., SH2 and SH3, frequently occur in signaling proteins. Src homology domains are often present in sites of tyrosine phosphorylation in receptor kinases. They subsequently bind to downstream targets and effector proteins.

Signaling Pathways: Relevance in Development

Insights into signal transduction pathways in humans have been derived in part through identification of human genes homologous to signal trans-duction genes present in other organisms. Insights have also been gained through identification of specific gene defects associated with particular developmental defects in humans.

Rasopathies

Signal conduction through the RAS MAP kinase signaling pathway plays key roles in early and late stage developmental processes(see Figure 3.1). Tartaglia and Gelb (2010) reported that dysregulation through heterozy-gous mutations in RAS proteins, RAS regulators and downstream effectors lead, in humans, to a group of clinically related disorders. The RAS MAP kinase signaling pathway plays role in regulation of cell cycle, cellular growth and differentiation. Rauen (2013) reviewed this pathway and the role of the different components within the pathway, in causation of the developmental defects known as Rasopathies. Rauen (2013) noted that although the seven disorders classified as Rasopathies exhibit a unique phenotype, they share overlapping features. These include cranio-facial

RAS SIGNALING PATHWAY

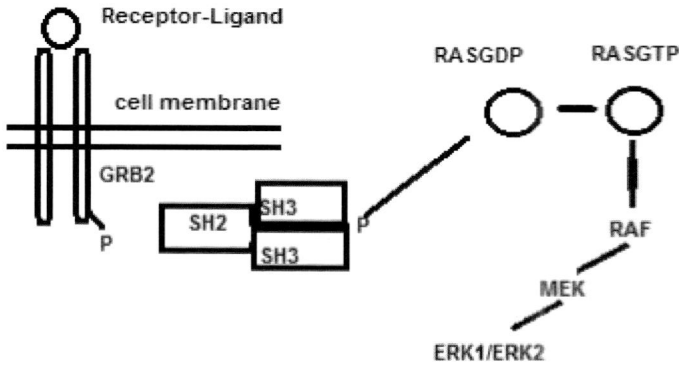

Fig. 3.1: RAS signaling.

dysmorphology, cardiac malformations, ocular defects, muscular-skeletal defects, hypotonia, and neurocognitive defects, cutaneous abnormalities that frequently involve pigment changes, and increased cancer risk. It is also interesting to note that a specific clinically defined disorder can arise as a result of defects in any one of a number of components in the RAS signaling pathway, e.g., Noonan syndrome.

It is important to note that congenital defects that constitute features of Rasopathies may result from activating or inactivating mutations in specific pathways leading to dysregulation of the signaling cascade.

Components of the pathway

Activity is initiated when a growth factor binds to a receptors and the receptor kinase triggers phosphorylation. This in turn triggers activation of a series of adaptor proteins including SH domain proteins, SHC, GRB2 and SOS and their activity lead to activation and to GTP bound forms of RAS including KRAS, HRAS and NRAS.

Activated RAS-GTP can trigger several downstream signaling cascades. One cascade includes sequential activation of RAF and the MAP kinase pathway. The later can then activate nuclear components including transcription factors.

RAS–RAF–MAP kinase pathway

1. Ligands that activate the receptors, e.g., the epidermal growth factor receptor.
2. Phosphorylation of cytoplasmic domains of receptor.
3. Adapter proteins GRB2, SH protein SOS1 dock with phosphorylated receptor.
4. RAS family proteins enter pathway (RAF, RAS, NRAS, KRAS and HRAS).
5. RAS-GDP is phosphorylated to generate RAS-GTP.
6. MAP kinase pathway components are activated.
7. MAP kinase pathway elements enter nucleus and impact transcription.
8. Negative regulatory elements include phosphatases, ubiquitin ligases including CBL (ring finger E3 ubiquitin ligase).

Syndromes due to mutations in RAS–MAP kinase pathway

Noonan syndrome: May arise from mutations in adapter molecular SOS, PTPN11, scaffold molecule SHOC2, or mutations in RAF, NRAS, KRAS and regulator CBL.

Neurofibromatosis: Defects in RAS-GTPase, results in increased active RAS.

Costello syndrome: Due to mutations in HRAS.

Legius syndrome: Mutations in SPRED1, (SH protein) active RAS suppressor.

Capillary malformation syndrome: Mutations in RASA1 (RASp21 protein activator).

Clinical manifestations in syndromes due to mutations in the RAS signaling pathway

Noonan syndrome results from mutation in NRAS, KRAS, RAF1 or mutation in SHOC2 that forms a scaffold that links RAS to downstream signal transducers. It may also arise as a result of mutation in the adaptor proteins SH domain proteins, GRB2, and SOS1 or as a result of mutations in negative regulators of the pathway including CBL that acts as an ubiquitin ligase and promotes degradation of specific pathway components.

In Noonan syndrome the facial features are unusual and include broad forehead, wide spaced eyes, down sloping palpebral fissures and posteriorly rotated ears. Cardiac defects frequently occur; growth is impaired and neurodevelopmental delay is common. Rauen (2013) reported that 50% of patients with Noonan syndrome have defects in PTPN11, a protein tyrosine phosphatase.

A Rasopathy known as capillary malformation arterio–venous syndrome, associated with capillary malformations and arterio–venous fistulas, results from inactivating mutations of the RASA1 protein that promotes RAS-GDP formation.

Neurofibromatosis is associated with abnormal freckling, cafe-au-lait spots, bone abnormalities, ocular lesions (Lisch nodules); some patients have vascular abnormalities and tumors may arise in peripheral nerves. This syndrome arises as a result of defects in the NF1 neurofibromin protein that normally functions as a RAS-GTPASE to reduce RAS signaling through the formation of RAS-GDP.

Costello syndrome due to defects in the *HRAS* gene has some manifestations that overlap with other Rasopathies and in addition this syndrome has some unique features. These include soft skin, excessive wrinkling, redundant skin, cutaneous papillomas, gastro-intestinal malformations, failure to thrive, and hypertrophic cardiomyopathy.

Legius syndrome has features of mild neurofibromatosis, cafe-au-lait spots and macrocephaly are the main features, and neurofibromas are usually absent. It is due to mutation in an SH domains protein encoded by the SPRED1 locus (Brems *et al.*, 2012).

Mosaic rasopathies

Specific malformation syndromes result from mosaicism for mutations in the RAS signaling pathway components. Mosaicism results from post-zygotic mutations in a specific gene leading to the presence of more than one cell population in an individual, one population with the mutation and another without the mutation. Mosaic mutations in the neurofibromin gene (*NF*1) lead to segmental lesions (Hafner and Grosser, 2013). Mosaic *NF1* mutations lead to syndromes characterized by sebaceous nevi or keratinocyte epidermal nevi with extra-cutaneous manifestations including ocular, cerebral and skeletal defects in patients with mosaic mutations in *HRAS* or *KRAS*.

Transforming Growth Factor beta (TGFbeta) Signaling

TGFbeta signaling is initiated by ligands that bind to a multi-component receptor complex including type I and type II receptor serine threonine kinases. Massague and Xi (2012) reported that type II receptor subunits phosphorylate and activate type I receptors. A large family of TGFbeta-like receptors has been identified. TGFbeta-like receptors often form complexes with TGFR1 and TGFRII. TGFbeta-like receptors may independently react with different ligands, e.g., endoglin, activin, bone morphogenetic protein (BMP). (See Figure 3.2.)

TGFbeta, bone morphogentic protein and activins are evolutionarily conserved cell secreted cytokines that play roles in development and

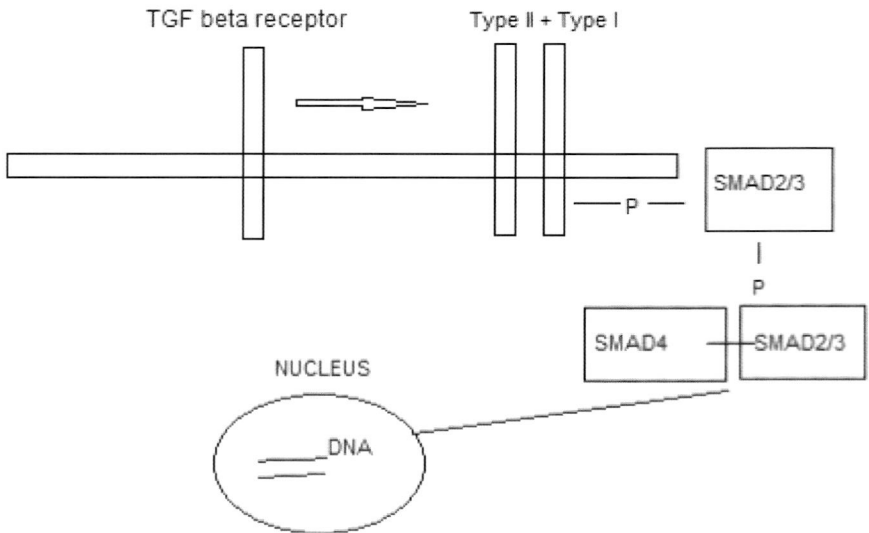

Fig. 3.2:   TGFbeta signaling.

throughout life. Abnormalities in signaling processes related to these molecules lead to a number of different diseases,

Regulation of TGFbeta synthesis, processing and storage

TGFbeta proteins are synthesized as precursor molecules that are then proteolytically processed. This processing includes removal of the signal peptide from pre-pro TGFbeta protein to generate pro TGFbeta protein. This molecule then undergoes dimerization. The pro–TGFbeta dimer is then cleaved by the protease Furin to generate a C terminal mature peptide and an N terminal remnant, the latency associated peptide (LAP). Both of these peptides remain in a complex that associates with a binding protein LTBP to form the large latency complex LTC. The LTBP can bind with a number of different molecules including fibrillin and other components in the extra-cellular matrix. The extra-cellular matrix plays an important role in storing and processing TGFbeta protein (Rifkin and Todorovic, 2010). It is important to emphasize that binding of TGFbeta mature peptide to LTBP latency binding protein enhances its interaction with fibrillin in the extra-cellular matrix.

Activation of TGFbeta requires its liberation from the extracellular matrix and microfibrils. Specific proteases, including matrix metalloproteases play roles in cleavage. The mechanisms of release of TGFbeta from the latency associated protein differ in different cell types. In some tissues integrins play roles; in other tissues metalloproteins are involved.

Released TGFbeta then binds to receptors, TGFbeta RII and then TGFbetaRI. TGFbeta receptors are particularly abundant in endothelial cells and in vascular smooth muscle cells. It is now known that TGFbeta signals through several pathways, including canonical and non-canonical pathways and through pathways that involve the activin and endoglin receptors.

TGFbeta canonical pathway

In the canonical pathway, TGF receptor activation leads to phosphorylation downstream of SMAD2 and SMAD3. These two proteins then form a complex with SMAD4. This complex is then transferred to the nucleus where it interacts with SMAD binding element and with DNA binding transcription factors to trigger expression of specific genes.

Non-canonical pathway of TGF signaling

In the non-canonical pathway of TGF signaling the activated TGFRII–TGFRI complex bypasses SMAD activation. Receptor phosphorylation leads to phosphorylation of other downstream signal transduction pathways including the SOS, GRB2 RAS–MAPkinase pathway. In addition, activated TGF receptors can signal through the phospho-inositide-3-kinase pathway and RAS and RHO and through Jun-kinase (JNK) or through NFkappa B pathways.

TGFbeta can also signal through activin receptor like kinase ACVRL1 and the co-receptor endoglin. Downstream signaling through SMADS then occurs. Activin and endoglin receptors are expressed on vascular endothelial cell surfaces (Fernandez *et al.*, 2006). Mutations in activin receptor, in the endoglin receptors and mutations in SMAD4 have been found to be responsible for arterio–venous malformations that occur in hemorrhagic telangiectasia also known as Osler Weber Rendu syndrome, an autosomal dominant disorder.

Control of levels of TGFbeta and TGFbeta-like receptors

There is evidence that ubiquitylation is a major regulatory mechanism. A specific ubiquitin ligase SMURF promotes polyubiquitylation of TGFRI. The recruitment of SMURF to receptors is facilitated by SMAD7(Bizet *et al.*, 2012). However, the action of SMURF is counteracted by other proteins including TRAF that de-ubiquitylates receptors (Moustakas and Heldin, 2013).

Marfan syndrome fibrillin and TGFbeta signaling

Mutations that impact structure or synthesis of fibrillin 1 lead to Marfan syndrome, an autosomal dominant disorder. Features of this disorder include long hands and feet (arachnodactyly), altered ratios of trunk to limbs, pectus excavatum, lens abnormalities (ectopia lentis), lax joints and high arched palate. The most clinically significant defects include aortic root dilation, ascending aorta aneurysm, valve prolapse and regurgitation. Aneurysms may occur in other blood vessels. There is also a tendency toward aortic valve regurgitation.

A key downstream effect of fibrillin defects or deficiency is impaired sequestration and binding of TGFbeta in the extra-cellular matrix, and increased TGFbeta and SMAD signaling. Losartan treatment reduces the excess TGFbeta signaling and improves the prognosis of Marfan syndrome (Dietz, 2014).

There is extensive overlap in pathology between Marfan syndrome and Loeys Dietz syndrome (LDS). LDS is frequently caused by loss of function mutations in the receptors TGFRII or TGFRI.

SMAD3 mutations have been associated with osteochondritis dessicans, mild hypertelorism, abnormal uvula and aortic aneurysms. It is therefore clear that the TGFbeta signaling pathway impacts development of the skeleton and the cardio–vascular system.

Hedgehog Signaling Pathway and Development

The sonic hedgehog (*SHH*) gene was first described in Drosophila and the human homolog of this gene was subsequently mapped to human chromosome 7q36. Following mapping of the developmental defect holoprosencephaly to this chromosome region, the *SHH* gene was investigated as a candidate gene for this condition. *SHH* gene mutations or structural defects were found to play roles in abnormalities of cranio–facial development and intensive studies of the SHH signaling pathway were initiated. Two SHH related genes were isolated in vertebrates and these were designated as desert hedgehog (*DHH*) and Indian hedgehog (*IHH*). These two genes have a much more restricted pattern of tissue expression. IHH protein plays an important role in bone development. DHH is primarily expressed in testes.

Varjosalo and Taipale (2008) reviewed the hedgehog signaling pathway. SHHencodes a protein that acts as a ligand for a cell surface receptor that triggers a downstream signaling cascade. SHH is expressed from early embryogenesis on; *SHH* gene expression is regulated by multiple enhancer elements. Several enhancers are located far upstream of the *SHH* transcription start sites while others are located in introns within introns of the *SHH* gene. One enhancer that is located approximately 1Mb upstream of the *SHH* promoter was found to be mutated in cases of congenital limb abnormalities with preaxial polydactyly.

SHH protein maturation involves removal of the signal peptide from the 45 kilodalton precursor protein that then undergoes auto-catalytic processing that involves cleavage between glycine and cysteine, this cleavage results in a thioester and the sulphate is then displaced by cholesterol. The protein is cleaved into a C-terminal domain that is non-signaling and a 19 kilodalton N terminal signaling domain. Lipid molecules are then added to the N terminal domain (Porter *et al.* 1996). The SHH signaling moiety occurs as a single molecule or as multimers.

SHH receptor

The SHH receptor is Patched 1 (PTCH1). When the SHH ligand is not present, this receptor is bound by an inhibitory molecule smoothened (SMO). When SHH is present, the inhibition of SMO is released and SHH binds to the PTCH1 receptor and to co-receptors. The primary cilia that protrude from cell surfaces act as the sites for SHH signaling. Binding of SHH to the PTCH1 receptor leads to processing and activation of GLI transcription factors through phosphorylation.

Developmental defects that result from loss of hedgehog signaling include holoprosencephaly, cranio–facial defects and skeletal malformations. It is important to note that mutations in at least 12 different genes can lead to holoprosencephaly. This condition results from failed or incomplete separation of the forebrain. Solomon *et al.* (2012) reported that *SHH* mutations account for approximately 12% of cases of holoprosencephaly. They reported that most of the *SHH* mutations are family or individual specific. Manifestations occur in homozygous and heterozygous mutation bearing individuals.

The GAS1 protein interacts directly with SHH and is a positive regulator of SHH activity. Pineda-Alvarez *et al.* (2012) reported that GAS1 mutations lead to holoprosencephaly. The GAS1 gene encodes a cell surface receptor that can act as a co-receptor for SHH. These investigators reported that defects in Nodal pathway signaling can lead to holoprosencephaly. Savastano *et al.* (2014) reported other genes and protein mutations that lead to holoprosencephaly; these include mutations in the homeobox transcription factor SIX3 and the transcriptional repressor ZIC2.

Non-canonical hedgehog signaling pathway

In the canonical hedgehog signaling pathway, the repressive effect of SMO on receptor PTCH1 is released upon binding of the hedgehog ligand to the receptor. This then leads to signals through the C-terminal end of SMO to activate the GLI transcription factors.

There is now evidence for a non-canonical hedgehog signaling pathway that does not involve GLI. Following hedgehog signaling, SMO activation leads to stimulation of small GTPs, RHOA and RAC1 and this stimulation leads to actin cytoskeletal changes. This non-canonical pathway is particularly important in endothelial cells that apparently lack GLI transcriptional activity (Polizio *et al.*, 2011).

Fibroblast Growth Factors (FGFs)

FGFs are polypeptides that are essential for many processes in embryonic development. They are also essential in post-natal life and play roles in inter-cell signaling as hormones and in metabolism (Itoh and Ornitz, 2011). (See Figure 3.3.)

In mammals there are 22 members of the FGF family and are often classified according to their functions as paracrine, endocrine and

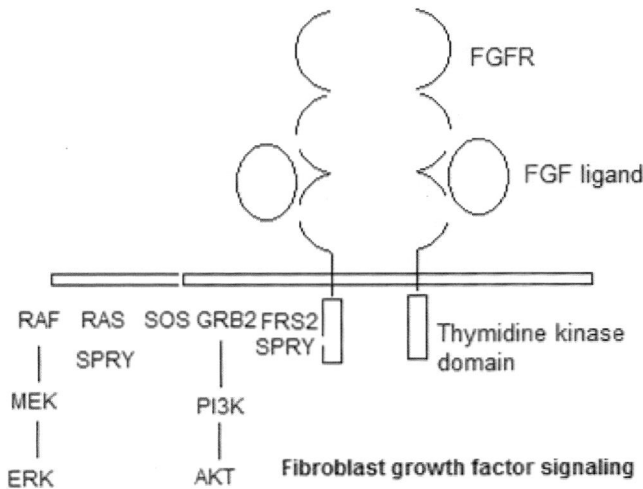

Fig. 3.3: FGF signaling.

intracrine. The paracrine FGFs bind to specific tyrosine kinase receptors on nearby cells. Endocrine FGFs exert effects following transport over long distances. Intracrine FGFs are intra-cellular molecules; they are not secreted and exert their effects through voltage gated sodium channels and with neuronal MAP kinase scaffold proteins.

The human FGF proteins range in length between 150–300 amino acids and have conserved cores of 120 amino acids. In the different forms of FGF, amino acids in the core show 30–60 amino acid identities.

Mason (2007) reported that FGFs play critical roles in neural crest development and in downstream development of neural crest derived cells. These include cranial neurons and glia, sensory neurons and glia, components of the sympathetic nervous system and adrenal axis, pigment cells, connective tissue, cartilage and bone.

FGF Receptors (FGFRs) Structure and Function

FGFR proteins are produced by at least four genes in humans. Alternate splicing of FGFR transcripts increases the number of FGFR proteins. The FGFRs have an extra-cellular region composed of three immunoglobulin-like domains. Immunoglobulin-like domain 3 is essential for ligand binding. In the case of paracrine, FGF heparin-like molecules are necessary to stabilize the ligand receptor interactions.

FGFRs are receptor tyrosine kinases with an extra-cellular ligand binding domain and a cytoplasmic domain that contains the tyrosine kinase core. The cytoplasmic region of the receptor also binds regulatory elements. Regulatory elements that interact directly with the cytoplasmic FGFRs include SHP2 (a tyrosine phosphatase), FRS2 (FGF substrate), SOS (guanine nucleotide exchange factor), SPRY (FGF antagonist) and GRB2 (growth factor receptor bound protein). FRS2 contains docking sites for other proteins including SHP2. FGFR activation then activates a number of downstream signaling pathways. These include mainly the RAS–RAF–MAP kinase pathway. However the phosphoinositol and AKT, and phospholipase C and protein kinase C pathways can also be activated.

Human developmental defects and FGFR

In 1994, Shiang *et al.* reported that activating mutations of FGFR3 led to achondroplasia dwarfism associated with skeletal malformations. Shortly, thereafter mutations in several different FGFRs, FGFR1, FGFR2, and FGFR3 were shown to be responsible for conditions characterized by cranial synostosis (early fusion of cranial bones) and skeletal abnormalities. These condition included Crouzon syndrome, Jackson–Weiss syndrome, Aperts syndrome and Pfeiffer syndrome. In these conditions, affected individuals are usually heterozygous for mutation in one of the FGFR genes.

Multiple different FGFR3 mutations were found to lead to hypochondroplasia-type dwarfism. FGFR3 mutations were shown to lead to inhibition of chondrocyte growth. Adequate bone growth needs to be preceded by adequate chondrocyte proliferation and differentiation.

Mason (2007) emphasized the important role of FGF signaling in nervous system development. It is important to note that molecules other than FGF also serve as ligands for FGFRs. These included cadherins, anosmin and neural cell adhesion molecule (NCAM).

Kallman syndrome is a developmental defect associated with underdevelopment of the olfactory bulb, anosmia, and sometimes with deafness and cleft-lip and palate. Specific FGFR1 and FGFR2 mutations lead to Kallman syndrome.

WNT Signaling

WNT genes encode a protein that is a secreted ligand that plays key roles in development. Van Amerongen and Nusse (2009) emphasized that both ligands and receptors in the WNT signaling pathway are members of large multigene families. There are 19 Wnt encoding genes in vertebrates. Although the signaling pathway that involves WNT beta-catenin is best known, they emphasized that WNT signals through other pathways that do not involve beta-catenin as the intra-cellular messenger.

Frizzled is the best known transmembrane receptor for WNT. There are 10 different frizzled receptors in humans. However there is now evidence that WNT may signal through other transmembrane receptors including RORA orphan receptor and RYK (receptor-like tyrosine kinase). In addition there is evidence that WNT signaling through frizzled may pass directly to RHO, RAC and JNK downstream signaling. The WNT pathways that do not involve beta-catenin are sometimes referred to as the non-canonical pathways. The non-canonical pathways can also include signaling through frizzled with co-activators LRP (low density lipoprotein co-receptor) and disheveled (DSH), through RAC, RHO to JNK or signaling through frizzled to phospholipase C and phosphatidyl inositol pathways. Signaling through frizzled most commonly involves recruitment of the low-density lipoprotein co-receptor and other co-factors including RSPO (R-spondin), DKK3 and Kremen 1(transmembrane protein).

Van Amerongen and Nusse (2009) emphasized that many different interactions are involved in determining the downstream effects of WNT signaling. They postulated that there is likely extensive cross talk between ligands, receptors, co-receptors and intra-cellular messengers. They noted that in organisms with different cell types and complex tissues, it might be less useful to continue to think of signal transduction pathways as linear cascades.

WNT Signaling in Development

There is evidence that WNT signaling pathways are involved in early and late developmental processes including cell proliferation, cell fate decisions and migration in limb, heart and neural development (Croce and McClay, 2008).

WNT plays an important role in skeletal systems; WNT expressed in bone marrow and hematopoietic progenitors was shown to impact development of osteocytes. There is evidence from studies in mice for the role of WNT in cross talk between hematopoietic and osteoblast lineages. In humans specific mutations in WNT1 have been identified in families with recessive osteogenesis imperfecta and in families with early onset osteoporosis (Laine *et al.*, 2013).

WNT also plays important roles in eye development. Specific mutations in the lipoprotein receptor protein LRP5 and in Frizzled4 receptor lead to specific eye conditions. These conditions are characterized by abnormal vascularization of the peripheral retina and the formation of intra-ocular fibro-vascular masses. Similar manifestations can be seen in patients with defects in the gene that encodes the Norrin protein. Norrin protein was first identified as defective in patients with Norrie disease. It is now clear that Norrin acts as a WNT ligand and is involved in the induction of beta-catenin signaling. Nikopoulos *et al.* (2010) reported that the pathology in these diseases is due to abnormal vascularization leading to an avascular peripheral retinal zone.

Nikopoulos *et al.* (2010) reported that LRP5 mutations are also associated with bone abnormalities including osteoporosis, and pseudoglioma syndrome that manifests in early childhood. Another bone abnormality associated with LRP5 mutation is endosteal hyperostosis and high bone mass.

Nuclear Receptor Signaling, Example Retinoic Acid Signaling Pathway

The retinoic acid signaling pathway differs from other pathways. In the retinoic acid signaling pathway the ligand does not act with cell surface receptors to transduce signaling. Retinoic acid enters the nucleus where it interacts with a nuclear receptor to impact gene expression (Duester, 2008).

Nuclear receptors are proteins in the nucleus that bind to specific sites in DNA. The unique property of these proteins is that when they bind to specific ligand, the ligand receptor complex binds to DNA and activates gene expression. Ligands for nuclear receptors include lipophilic molecules, specific fatty acids, retinoic acid, vitamin D and hormones including steroid and thyroid hormones.

Retinoic acid is derived from vitamin A, retinol, through a series of transformations that include conversion of retinol to retinaldehyde through the action of retinol dehydrogenases and alcohol dehydrogenases. Retinaldehyde is then converted to retinoic acid through the action of retinaldehyde dehydrogenases; degradation of retinoic acid is achieved through activity of cytochrome oxidases, e.g., CYP26A1and CYP26C1.

Duester (2008) presented evidence that the enzyme encoded by alcohol dehydrogenase functions primarily to reduce quantities of excess retinol.

Retinol binding proteins exist in the serum and cells. Retinol is stored in the liver and is transported in the circulation bound to the retinol binding protein RBP4. The protein encoded by the *STRA*6 (stimulated by retinoic acid) gene carries retinol into cells. Free retinol is used for synthesis of retinoic acid.

Nuclear receptors that bind all trans retinoic acid include receptors RAR alpha, RAR beta and RAR gamma. The RXR (alpha, beta and gamma) nuclear receptors bind a specific form of retinoic acid, 9-cis retinoic acid.

Duester (2008) reported that studies in mice revealed that retinoic acid is specific to developmental processes in the posterior neuro-ectoderm and in the foregut endoderm that gives rise to the pancreas and lung, it is also likely a signal to the trunk mesoderm. There is evidence that retinoic acid acts *via* nuclear receptors to target specific genes. Specific HOX genes, *Pax*1, *Pitx*1 and *Ret* are upregulated while *Fgf*8 is down regulated.

Retinoic acid receptors RARA, RARB, RARG and RXR can heterodimerize and can bind to response elements in genes (RARE). Binding of the ligand retinoic acid to the DNA bound receptor results in release of the co-repressor and binding of co-activators.

Specificity of Signaling Processes in Development

A puzzling fact is that a specific signaling pathway leads to different developmental outcomes in different tissues. Van Amerongen and Nusse (2009) emphasized that there is likely extensive cross-talk between different ligand receptors, co-receptors and intra-cellular messengers They noted that in organisms with different cell types and complex tissues it might be less useful to continue to think of signal transduction pathways as linear cascades.

Different results of activation of a specific signaling pathway may lie in the fact that a specific gene, e.g., a specific ligand or a specific receptor gene, gives rise to a series of different isoforms, i.e., different protein products. Determination of the specific product that is produced from a specific gene depends in part on the environment in which the gene expression takes place, on the additional products that are present in the specific environment and on the regulatory elements that are operating at the time of transcription.

Considered below are the complexities of vascular endothelial growth factor signaling determined through transcription initiation from different promoters, generation of different splice forms and differences in translation and mRNA stability.

Vascular Endothelial Growth Factor (VEGF) Signaling Pathway

The signaling pathway that involves VEGF as ligand and VEGFR as receptor serves as an example of the complexities and variation of transcription, post-transcriptional modification, and variations in translation and post-translational modification.

VEGF produced by endothelial cells, signals through two receptors, VEGFR1 and VEGFR2. Signaling is primarily through VEGFR2 that acts as a receptor kinase. Activated VEGFR2 recruits adaptor proteins SHC GRB2 and phospholipase 2 and the downstream effect is through the RAF–MAP kinase pathway (MEK, ERK).

VEGF is essential for endothelial cell proliferation and migration. Wagner and Siddique (2007) reported that in cardiac development, VEGF signaling plays essential roles in endocardial cushion formation and subsequent formation of cardiac valves. In adult tissues endothelial cells are protected from apoptosis by VEGF.

Arcondeguy *et al.* (2013) published an extensive review of VEGF transcription, mRNA stability and translation and analysis of VEGFA expression levels. The *VEGFA* gene maps to human chromosome 6p21.1 and it encompasses approximately 14kb. In human, the *VEGFA* promoter extends over 2.36kb and has binding sites for a number of different transcription regulators.

Hypoxia serves as an important stimulus for VEGFA expression and for hypoxia induced factors (HIF). HIFs interact with a hypoxia responsive element (HRE) in the 5′ flanking region of *VEGFA*.

VEGFA has an alternative promoter which is located 633 nucleotides downstream of the common transcription start site. *VEGFA* has eight exons and alternative splicing occurs. The four N-terminal exons are constitutive and are present in all isoforms and the other four exons are alternative. Alternative splicing generates nine different isoforms. VEGFA transcript isoforms that lack exons 6 and 7, also lack heparin binding capacity and readily diffuse out of cells. VEGFA isoforms 165 and 189 include exons 7 and 8 and bind heparin sulfate and aggregate on the cell surface. In addition alternate splicing can occur within exon 8 leading the VEGFA isoforms with different carboxy terminals, Cys–Arg–Lys–Pro–Arg–Arg for exon 8a, and Ser–Leu–Thr–Arg–Lys–Asp for exon 8b. These two isoforms have different interaction capacities.

What factors lead to generation of specific splice forms? Arcondeguy *et al.* (2013) noted that one factor is the activity of an exonic splicing silencer sequence that likely binds to specific regulatory sequences.

Messenger RNA stability

Arcondeguy *et al.* (2013) reported that there are two major polyadenylation sites in *VEGFA* gene. Variations in VEGFA mRNA stability are impacted by environmental changes in the cell and also by AU rich elements (AREs) that occur in the 3′ non-coding region upstream of the polyadenylation signal. Specific factors that bind to the AREs may promote degradation of mRNA while other AU binding factors promote mRNA stability.

There is evidence for transcriptional regulation of VEGFA expression. Capping of primary mRNA transcripts is a process that subsequently impacts transcription and translation. Capping involves addition of the nucleoside 7-methylguanosine that is linked to the first 5′nucleotide of the transcribed sequence. Although most mRNA transcripts undergo capping there are transcripts that do not. Those transcripts frequently have long 5′untranslated regions that contain certain specific sequence elements, internal ribosomal entry points (IREs). There is evidence that in the presence of IREs, translation may start upstream from the usual AUG start site. In VEGFA translation may start at a CUG codon leading to translation of a product that is 206 amino acids longer than the regular protein. This LVEGFA protein subsequently undergoes protein cleavage to generate NVEGFA. There is some evidence that NVEGFA may have specific functions.

There is a conserved open reading frame (ORF) upstream of the *VEGFA* gene and evidence that this ORF plays a role in regulation of VEGFA mRNA translation. The ORF interacts with ribosomes and limits access of ribosomes to the main AUG translation initiation site.

At the 3′ end of the *VEGFA* gene there are sequence elements that can bind proteins and this impacts mRNA folding. These sequence elements are referred to as Riboswitches. For VEGFA, Arcondeguy *et al.* (2013) reported that environmental factors, e.g., hypoxia, influence which proteins bind to riboswitch elements. Riboregulation is turning out to be an interesting aspect of eukaryotic gene expression.

Specific microRNAs that bind to VEGFA transcripts have been identified. They are likely to impact translation.

VEGF Receptors

VEGF signals through two receptors, however signaling occur primarily through VEGFR2. Saito *et al.* (2013) published evidence that a soluble form of VEGFR1 sequesters VEGFA protein and acts as a potent anti-angiogenic factor. They reported that the soluble form of VEGFR1 is a splice form that undergoes premature polyadenylation. They determined that VEGFA up-regulates soluble VEGFR1 expression through interaction with a VEGFA response element in intron13 of VEGFR1. They emphasized that spatial and temporal regulation of angiogenesis is vital during and after embryonic development. The VEGFA–VEGFR1 interaction serves to inhibit angiogenesis.

CHAPTER 4

PLURIPOTENCY TO DIFFERENTIATION

Insights into the Molecular Biology of Development through Stem Cell Studies

Pluripotent stem cells have the capacity for indefinite self-renewal and the capacity to differentiate to all tissues of the body. Analysis of the factors that lead to lineage commitment and differentiation provide insights into mechanisms of mammalian embryogenesis (Pera and Tam, 2010).

Embryonic Stem Cells (ESCs) Development

Research on teratomas and teratocarcinomas can be considered as key starting points for stem cell research. These tumors occur in humans; however they also occur in mice and mouse teratocarcinomas (also known as embryonal carcinomas) were utilized for many research studies. These tumors are unique in that they contain a variety of different cell and tissue types with different degrees of differentiation. Kleinsmith and Pierce (1964) reported that a single cell isolated from a mouse teratocarcinoma and then transplanted to the peritoneal cavity of another mouse of the same strain could give rise to all the cell types present in the original tera-tocarcinoma. Studies were undertaken to identify specific protein markers that were uniquely expressed on the pluripotent mouse embryonal carci-noma cells. In 1978, Strickland and Mahdavi reported that retinoic acid induced differentiation in cultured embryonal carcinoma cells.

In 1981, Evans and Kaufman reported that they had established cul-tures of pluripotent stem cells from mouse blastocysts. In the same year, Gail Martin reported establishing a pluripotent stem cell line from the inner cell mass of late stage blastocysts derived from pre-implantation mouse embryos.

In 1998, Thomson and co-workers reported success in deriving pluri-potent cell lines from human blastocysts. The inner cell mass isolated from blastocysts underwent undifferentiated proliferation for at least eight months. Thomson *et al.* (1998) reported that these human ESCs produced high levels of telomerase. Telomerase maintained chromosome telomeres and this was important for ongoing cell replication. They noted further that the human ESCs expressed specific embryonic antigens and the embryonic (placental) form of alkaline phosphatase. When these embryonic cells were injected into mice, they produced teratomas that contained endoderm, mesoderm and ectodermal components.

Byrne *et al.* (2003) reported that nuclei of adult human lymphoblasts injected into enucleated Xenopus oocytes were reprogrammed. In the reprogrammed cells the pluripotency marker OCT4 (homeobox transcription factor) was strongly expressed. In 2005, Simonsson and Gurdon reported that following somatic cell nuclear transfer to an enucleated oocyte, genes that were active in the differentiated somatic cell nuclei prior to transfer, were silenced following transfer. Genes that were silenced in the somatic nuclei became actively expressed following transfer of those nuclei to an oocyte. They emphasized that epigenetic changes, such as chromatin remodeling and DNA demethylation were involved in the activation of expression of OCT4 in the reprogrammed cells.

Studies by Chambers and Smith (2004) determined that pluripotency in ESCs was dependent on the homeobox transcription factors OCT4, SOX2 and NANOG.

Induced Pluripotent Stem Cells

In 2006, Takahashi and Yamanaka reported that cells from differentiated tissues could be reprogrammed to produce pluripotent cells, using a limited number of transcription factors. From such pluripotent cells, a variety of different tissues can be produced. The factors that induced pluripotency, OCT4, SOX2, KIF4 (kinesin family member) and cMYC (nuclear phosphoprotein) were introduced in a specific plasmid vector. Following transformation to pluripotency, a variety of different factors were used to induce differentiation to specific cell types.

John Gurdon and Shinya Yamanaka were awarded the 2012 Nobel Prize in Physiology and Medicine for the discovery that mature cells can be reprogrammed to become pluripotent.

Through understanding the molecular mechanism of stem cell self-renewal and pluripotency, we will gain understanding of development and insight into developmental defects.

Analysis of Regulators of Gene Expression

Young (2011) emphasized the importance of analysis of the functions of regulators of gene expression programs. Regulators may induce gene silencing; they may also lead to genes being in a poised state for activation, or being repressed.

Early insights into regulation of gene expression came through detailed analysis of the lac operon in bacteria, first by Jacob Monod (1961). In reviewing these studies, Young (2011) noted that in the absence of lactose, a specific repressor, bound to the lac operator inhibited transcription of the genes involved in lactose metabolism. In the presence of lactose, expression of genes involved in lactose metabolism was activated through loss of the repressor, binding of transcription activating factors and recruitment of RNA polymerase.

Young (2011) noted that the lac operon model provides a foundation for understanding gene expression. Gene expression is controlled through specific elements that bind to DNA. There is now evidence that transcription factors represent 10% of coding genes in the genome. They bind to elements proximal and distal to genes. Active gene expression involves recruitment of transcription factors, release of factors that cause pausing and recruitment of RNA polymerase.

Transcription factor auto-regulatory loop in stem cells

Young (2011) emphasized two key concepts in the control of the ESC state. Firstly, the transcription factors form an auto-regulatory loop; they work together to positively regulate their own promoters. In addition to activating expression of genes involved in maintaining pluripotency, they repress expression of transcription factors that lead to differentiation. Repression of expression of the differentiation genes depends in part on the binding of these factors to chromatin regulators, e.g., SetDB1 and Polycomb complexes associated with histone 3 lysine K9 trimethylated

(H3K9me3). Other important chromatin regulators are CHD1 and CHD7 (chromodomain helicases).

There is evidence that specific transcription factors recruit RNA polymerase II to promoters. In some cases there is evidence that recruited RNA polymerase II requires the presence of kinases to activate expression.

Young (2011) reviewed enhancers and enhanceosomes. Enhancers bind multiple transcription factors and co-factor complexes. There is evidence for multiple seemingly redundant enhancers and that this redundancy may contribute to phenotypic robustness. Enhancers also bind OCT4/SOX2 and NANOG.

Co activators and co-repressors of gene expression in stem cells

Young (2011) reviewed DNA binding co-factors including co-activators and co-repressors and chromatin regulators that are important in regulation of gene expression in stem cells. Specific transcription factors bind to co-activator proteins. There is evidence that the Mediator complex and Cohesin facilitate looping and linkage of transcription factor bound enhancers to promoters and impact gene expression. Mediator is also responsive to signaling. Other co-repressor complexes involved in control of embryonic cell gene expression include DAX1 (nuclear receptor regulator of transcription), CNOT3 (member of transcription regulator complex) and TRIM28 (tripartite motif containing repressor).

Chromatin regulators and histone modifying enzymes that impact the ESC state include the Polycomb complex (PCG) and the Trithorax complex (TRXG) and methyltransferase including SETDB1 and multiple histone H3 lysine methyltransferases.

Transition to differentiation states involves loss of OCT4 and NANOG through a variety of different mechanisms that include proteolytic destruction. Other mechanisms that promote differentiation include silencing of repressors of expression; modification or chromatin bound subunit complexes, e.g., Mediator and increased expression of microRNAs.

Mediator and transcription elongation

Conaway and Conaway (2013) described Mediator as enormous multisubunit complex apparently exclusive to eukaryotes. Mediator has an array of surfaces that can bind to RNA polymerase and to DNA transcription initiation complexes and to general transcription factors. Mediator is described as an RNA polymerase II co-regulator complex.

General transcription factors are the minimum set of transcription factors required by RNA polymerase II. They include TFllB TFllD, TFllE, TFllF and TFllH. At active promoters, RNA polymerase II and ribonucleotide triphosphates assemble.

Alternate splicing in stem cells

Han *et al.* (2013) analyzed differences in alternate splicing between pluripotent and differentiated stem cells. They reported differences in the levels of splice regulators MBNL1 and MBNL2 (muscleblind-like RNA binding proteins). Differentiation was associated with increased levels of MBNL1 and MBNL2 and decreased expression of FOXP1 (Forkhead transcription factor). There are reports indicating that epigenetics and alternate splicing are linked.

Han *et al.* (2013) reported that over-expression of MBNL1 and MBNL2 promoted differentiation. In the presence of MBNL1 and MBNL2, alternate splicing of FOXP1 occurred and this played a role in pluripotency. Their analysis revealed MBNL1 binding motifs in FOXP1 upstream of exons impacted exon inclusion.

Alternate splicing in a specific gene leads to differential inclusion of exons in the mature mRNA transcript and to different protein isoforms derived from a specific gene. Alternate splicing of the primary transcript of FOXP1 leads to generation of a protein with altered DNA binding properties.

Signaling factors for pluripotency

Differentiation is driven by specific signaling factors and the response to signaling is induction of transcription factors and induction of expression of specific genes. In addition, the state of the epigenome impacts response to signaling and transcription factor binding. Pera and Tam (2010) emphasized that specific signaling and transcription factor production required for differentiation are not always identical in mice and humans. They emphasized further that pluripotent stem cell populations are in fact heterogeneous in that within the IPS populations there are individual cells that differ from each other in their response to specific signals.

Pera and Tam (2010) reported that, particularly for the purposes of regenerative medicine, it is necessary to analyze the culture conditions and factors necessary for maintaining pluripotency.

The *NANOG* homeobox gene is a key factor and expression of this gene is important in maintaining pluripotency of human stem cells. Key factors for maintaining pluripotency in human ESCs include molecules in the TGFbeta signaling pathway, particularly activin, nodal and TGFbeta. These factors stimulate production of SMAD2 and SMAD3 that bind to the promoter of *NANOG* and stimulate its expression. Differentiation of ESCs is suppressed through binding of SMAD1, SMAD5 and SMAD8. These transcription factors are expressed in response to signal from morphogenic proteins encoded by *BMP2, BMP4* and *BMP7* genes.

FGF2 (fibroblast growth factor 2 is also crucial for retaining pluripotency in human ESCs. FGF2 activates signaling through FGF receptors and downstream targets receptor tyrosine kinases in the MEK ERK2 cascade.

Multiprotein Complexes, Chromatin Insights into Pluripotency and Differentiation

During development specific gene expression programs determine cell proliferation, differentiation and generation of diverse cell and tissue types. These programs are determined through modulation of transcription factor production, through modification of chromatin and ultimately through modulation of gene expression.

Key multi-protein complexes involved in these processes include Polycomb and Trithorax developmental regulators. Polycomb complexes play roles in development through repression of expression of specific genes. Soshnikova (2011) reported that genes encoding transcription factors, signaling molecules and receptors constitute approximately half of the Polycomb gene targets. Polycomb binding is strongly associated with trimethylation of lysine 27 in Histone H3 (H3K27me3). However, many genes have bivalent domains that include H3K27me3 and the histone form associated with active genes, namely trimethylation of lysine 4 in histone H3 (H3K4me3).

Polycomb repression of gene expression likely plays important roles in maintenance of pluripotency and suppression of differentiation. Polycomb proteins (PCG) are recruited to specific sequence elements in DNA polycomb response elements. These response elements often occur near Cytosine–Guanine (CpG) rich promoters of genes. However, DNA segments encoding long non-protein coding RNA (lncRNA) have also been found to harbor PCG binding elements (Soshnikova, 2011).

Trithorax Complexes (TRXGs)

TRXGs are associated with actively expressed genes and with trimethylation of lysine 4 in histone H3 (H3K4me3). Schuettengruber *et al.* (2011) reported that in vertebrates, trithorax multi-protein complexes bind to DNA. Trithorax components participate not only in histone methylation but also in histone acetylation. In addition, TRXG components participate in chromatin remodeling. Remodeling comprises nucleosome sliding, nucleosome eviction and chromatin looping.

Schuettengruber *et al.* (2011) listed nine categories of TRXG proteins. There are proteins involved in methylation (e.g., MLL1–4) (also known as KMT2A, KMT2B, KMT2C and KMT2D), and proteins active in histone acetyltransferase activity (ASH1 and ASH2). ATP dependent chromatin remodeling complexes include BRM, BRG1. Other components involved in chromatin remodeling include CHD (chromodomain helicases) and WDR5 (WD repeat domain protein).

There is evidence that TRXG components play key roles in cell cycle regulation and the MLL (KMT) methylases are particularly important. MLL proteins also act in detecting DNA damage prior to cell division.

Schuettengruber *et al.* (2011) reported results of studies on adult stem cells, including hematopoietic stem cells and neural progenitor cells. They noted that promoters in neural stem cells are frequently bivalents, they retain both H3K27me3 and HK4me3 and they also retain pluripotency and capacity to differentiate. The promoters in hematopoietic stem cells are also bivalent.

RE1-Silencing Transcription Factor (REST) Neuron-Restrictive Silencer Factor (NRSF) Multipotent and Progenitor Stem Cells

Covey *et al.* (2012) analyzed multipotent stem cell and neural progenitor cells (NS/P). They determined that NS/P cells that lacked REST (RE1-silencing transcription factor) retained the capacity to differentiate to neurons and glia but had reduced self-renewal capacity. In REST minus NS/P cells there was upregulation of genes involved in neuronal and oligodendrocyte formation and myelination.

Covey *et al.* (2012) emphasized the importance of long-term maintenance of the neural stem cell and progenitor cell pool. They proposed that REST expression plays an important role in this maintenance.

Target genes that bind REST have a 23 base pair sequence element, the neuron-restrictive silencer element (NRSE), to which REST and its co-repressor COREST bind. In non-neuronal cells this binding suppresses gene expression. However, there is evidence that in stem cells the repressed genes are poised for reactivation.

Gao *et al.* (2011) reported that in adult brains, REST is required to maintain the adult stem cell pool and to co-ordinate stage specific differentiation. They emphasized the importance of maintaining the stem cell pool in the hippocampal sub-granular zone and in the sub-ventricular zone of the lateral ventricles. REST expression is very low in cortical neurons. However, there is evidence that it is somewhat higher in adult hippocampal granular zone and pyramidal neurons, and that it is up-regulated in response to ischemia and seizures.

Gao *et al.* (2011) determined through gene knockout studies in mice that REST expression and COREST and SIN3A (transcription regulator) are essential for maintenance of the neural stem cell pool. Their studies revealed that loss of REST promotes premature exit of stem cells from quiescence. They determined that REST repression of specific genes was particularly important. These genes included ASCL1 (transcription factor important in neuronal differentiation) and NEUROD1 (transcription factor).

Modeling Human Diseases and Stem Cells

In recent years, mice have increasingly been used to produce models of human disease. One difficulty that sometimes emerges when mouse models are used to investigate the effects of a specific human gene mutation is that the mouse develops a different constellation of symptoms (Tiscornia, Monserrat *et al.*, 2011 and Tiscornia, Vivas *et al.*, 2011). For example, when specific cystic fibrosis gene (CFTR) mutations that cause respiratory disease in humans, are introduced into mice, the mice die of intestinal obstruction before they develop respiratory symptoms. There is also evidence that medications that work well in curing a specific disease in mice fail to cure that disease in humans.

Induced pluripotent stem cells (iPSCs) are used to study the effects of specific gene mutations and their impact on developmental processes. The iPSCs derived from a specific patient can be used to study how a particular gene mutation impacts physiological processes and leads to disease. In addition, iPSCs are used in gene targeting and gene correction to provide sources of cells with normal function.

In several studies, pluripotent stem cell derived from fibroblasts of patients with cardiac arrhythmias were differentiated into cardiac myocytes and into ventricular, atrial or nodal myocytes, (Moretti *et al.*, 2010). In the derived myocytes, they demonstrated altered ion channel activation and deactivation, and reduction in current. Increased catecholamine sensitivity was also found in the derived myocytes and this sensitivity was reduced by pharmacological beta-blockers.

Coupling iPSCs with gene targeting, editing and mutation correction

Studies are frequently carried out in model organisms, particularly in mouse to establish the role of specific genes in developmental processes and to determine how genes impact phenotype. Genome editing was sometimes accomplished by random mutagenesis. Knockout techniques or gene inhibition techniques were also used. Induction of mutations or gene knockout were then followed by selective breeding of organisms with mutations.

More recently nuclease based gene and genome editing methods have been designed to induce mutations or to correct mutations, and to modify genome architecture. Factors involved in gene regulation may also be investigated through genome targeting. There is also evidence that nuclease based gene and genome modification methods may be optimized and applied to correct gene defects.

Gene editing requires introduction of a double stranded DNA break at a specific genomic site. Nucleases are used to introduce these breaks. In different systems of gene editing different mechanisms are used to bind nucleases to the target genomic site. Three well-studied nuclease systems have been in gene targeting strategies. The first system used involved zinc finger proteins that contain DNA binding motifs coupled to Fok1 restriction endonuclease. Zinc finger DNA binding motifs are rich in cysteine and histidine. Each distinct zinc finger recognizes three bases in DNA, for example, 18 bases can be recognized by six tandem linked zinc fingers. The endonuclease is linked to the zinc fingers but is separated from them by a few bases and can potentially cleave DNA. Double stranded DNA breaks occur when zinc fingers are bound to opposite strands of DNA and cleavage occurs between the strands. The double stranded break can be repaired by homology directed repair when appropriate sequences are provided. Repair can also be achieved through non-homologous end joining.

Talens (Tale based nucleases) are plant derived factors. Each Talen repeat is composed of 34 amino acids; within a repeat there are two variable amino acids that determine binding to one DNA base. The amino acid recognition elements for each specific nucleotide can be arranged to target a specific DNA sequence. Fok1 nucleases are inserted between the left and right arm of the Talens. Following cleavage of DNA, correction can be achieved by homologous recombination of non-homologous end joining.

In these techniques, plasmids are often used for zinc finger bound nuclease or Talen delivery and for delivery of repair DNA. However, in some cases delivery of specific nucleases was achieved by electroporation. In addition, there is evidence that single stranded oligonucleotide sequences can be used for repair (Chen *et al.*, 2011).

The most recently devised gene targeting systems, the CRISPR CAS 9 system is RNA based. The targeting features of this system are much simpler to construct than the protein entities required for targeting with zinc fingers or Talens.

CRISPR loci in bacteria are composed of repetitive elements interspersed with short (approximately 20 base pairs) unique sequence elements (protospacers). In the natural bacterial CRISPR system, the protospacers correspond to sequence of viral elements that previously infected the bacteria. Transcripts are derived that contain part of the CRISPR repeat and sequence corresponding to the protospacers. In addition, 20 nucleotides of sequences referred to as protospacers associated sequence (PAM) is required for correct targeting. This transcript referred to as crRNA sequence then hybridizes to a transactivating RNA (trac RNA). This fusion product then forms a complex with CAS9 nuclease. The protospacer sequence subsequently binds to the corresponding DNA in invading organisms.

Bioengineering systems produced include CAS9 nuclease bound to guide DNA which is comprised of crRNA and trac RNA. The RNA sequence corresponding to the genomic site to be targeted is placed 5′ to the PAM sequences.

Following cleavage of the genomic site by the targeted nucleases repair can be induced through non-homologous end joining, resulting in deletion at the target site. Repair can also be initiated by providing DNA segments and through homologous recombination (Sander and Joung, 2014).

CHAPTER 5

DEVELOPMENT AND GROWTH ABNORMALITIES

Environmental Factors

In considering the etiology of growth abnormalities and birth defects it is important to take into account consequences of deficiency of macro and micro-nutrients and the exposure to harmful substances in the environment.

Neuro-disability and nutrition during pregnancy and infancy

Kerac *et al.* (2014) reported that in humans, most neurodevelopment is completed during the first 1,000 days after conception, a period that includes intrauterine life and early infancy. Nutritional deficiency in pregnant women increases the risk in their offspring for cognitive delay and for neurobehavioral defects later in life. Particularly important nutrients during pregnancy and early infancy are folate, vitamin A, iodine, iron and zinc and long-chain polyunsaturated fatty acids.

Kerac *et al.* (2014) emphasized that it is important to consider the hazards of adolescent pregnancy and childbirth. Pregnancy in young adolescents has the risk of being associated with cephalo-pelvic disproportion, difficult labors and birth processes with associated fetal hazards. In addition young adolescents more frequently give birth to low birth weight infants who are at risk for hypoxia, jaundice, kern icterus and subsequent cerebral palsy.

Iodine deficiency

Iodine deficiency and adequate synthesis of thyroid hormone are essential for normal development and maturation of the central nervous system. Iodine deficiency during pregnancy and early infancy can lead to intellectual impairment and in some cases it is associated with deafness and other neurological deficits (Donald *et al.*, 2014).

Worldwide iodine deficiency is the single largest cause of preventable psychomotor impairment and hearing and speech defects. Zimmerman *et al.* (2008) reported that insufficient iodine intake is particularly a problem in South Asia and sub-Saharan Africa.

Neural tube defects

Adequate intake of folic acid during early pregnancy has repeatedly been shown to reduce the incidence of neural tube defects including anencephaly and spina bifida. However, epidemiological research reported by Bell and Oakley (2009) indicated that worldwide only 10% of the possible prevention of neural tube defects was achieved.

Reducing the Incidence and Impact of Birth Defects

In a survey published in 2008, Howson *et al.* (2008) reported that every year 7.9 million children are born into the world with a serious birth defect determined by genetic or partly genetic causes. They noted that in addition, 1 million more children are impacted by post-conception factors, including nutrient or micronutrient deficiencies, alcohol exposure or exposure to infectious agents. Howson *et al.* (2008) estimated of these children 3,2 million survive early infancy have life-long mental, auditory, visual or physical impairments. The impacts or birth defects are particularly problematic in low and middle-income countries.

Turnpenny and Ellard (2004) published evidence that in high-income countries 40% of birth defects were due to pre-conception conditions including 6% induced by chromosome abnormalities, single gene disorders 7.5%, 20–30% due to multifactorial disorders. For post-conception birth disorders in high-income countries 7–8 % of cases were attributed to teratogenic factors.

It is interesting to note that availability of complete sequence of the human genome and the implementation of high throughput sequencing techniques is leading to recognition of a growing number of underlying genomic and genetic factors involved in causing birth defects.

Fetal Alcohol Spectrum Disorders

Fetal alcohol spectrum disorders include complete or partial fetal alcohol syndrome, alcohol related birth defects and alcohol related neurodevelopmental defects. Fetal alcohol exposure can lead to craniofacial abnormalities, neuro-anatomic abnormalities, cardiovascular malformations, growth abnormalities and behavioral problems.

The craniofacial features include microcepahly, narrow palpebral fissures, epicanthal fold, flattened nasal bridge, smooth philtrum, thin upper lip and ear folds may be unusual.

Prenatal fetal alcohol exposure also predisposes to spontaneous abortion, premature birth with underweight newborns and to increased incidences of stillbirths.

Heller and Burd (2014) analyzed ethanol distribution and elimination from the fetal compartments following maternal alcohol consumption. They reported that a 140 pound women who has four standard drinks (14 grams of alcohol per drink) will take 8.5 h to reach a blood alcohol level of 0. At the peak alcohol concentration in the maternal blood the fetal blood has a similar concentration. However, elimination of alcohol from the fetus is much reduced in comparison to the mother.

Alcohol enters the fetus from the placenta through the cord vessels and through membranes. Also alcohol excreted in urine from the fetus enters the amniotic cavity. Prior to 20 weeks of gestation the amniotic fluid alcohol is partly reabsorbed by the fetal skin. Throughout much of pregnancy the fetus swallows amniotic fluid. In the intestine, the alcohol is partly reabsorbed into the circulation and liver. Ethanol breakdown is initiated by liver alcohol dehydrogenase. Heller and Burd reported that alcohol elimination rate from the fetus is approximately 3–4% of that in the mother.

Smith *et al.* (1971) reported developmental changes in the expression of alcohol dehydrogenase during fetal and post-natal development. They reported that ADH1 (ADH1A in new nomenclature) is the only liver ADH isozyme expressed in the first trimester of human life. The ADH2 locus (ADH1B in new nomenclature) is expressed after the first trimester. Low levels of ADH 2 (ADH1B) are expressed in lung throughout life, in fetal kidney and intestine the ADH3 enzyme (ADH1C) is expressed.

Polymorphic variants that impact enzyme activity occur in ADH2 and ADH3 (ADH1B and ADH1C) enzymes (Smith *et al.*, 1972).

There is a lack of information as to how alcohol exposure leads to the spectrum of abnormalities in fetal alcohol syndrome. Wang and Bieberich (2010) reported on the basis of studies of cultures of neural crest cell, that alcohol promotes ceramide accumulation in these cells. This ceramide accumulation promotes apoptosis of neural crest cells. These authors note that many of the tissues impacted in fetal alcohol syndrome contain neural crest derivatives.

A number of authors including Ungerer *et al.* (2013) and Khalid *et al.* (2014) have reported that epigenetic mechanisms, including aberrant DNA and histone methylation, were impacted by ethanol exposure.

Medications consumed by pregnant women that may lead to fetal developmental defects

Evidence has accumulated (Brent, 2004) indicating that the following medications consumed during pregnancy may predispose to birth defects: Aminopterin (methotrexate), specific anti-epilepsy medication, trimethadione, Valproic acid, Carbamazepine, anti-tuberculosis medication Isoniazid, retinoid (acutance), diethylstilbestrol, anti-thyroid medications and thalidomide.

Nicotine exposure

A number of adverse effects on development have been attributed to nicotine exposure during pregnancy. These include increased risk of spontaneous abortion, growth restriction, sudden infant death syndrome and behavioral defects (Shea and Steiner, 2008).

The underlying mechanisms for nicotine related birth defects might include the effect that nicotine has in promoting constriction of vasculature. In addition nicotine binds to receptors and leads to dysregulation of neurotransmitter systems.

Maternal diabetes and birth defects

Pre-pregnancy diabetes and gestational diabetes are associated with a slight increase in the risk of specific birth defects in particular neural tube defects and cardio-vascular defects (Corrigan *et al.*, 2009). Another key problem that results from poorly controlled diabetes during pregnancy is macrosomia of the fetus, and the difficulties associated with the delivery of a very large fetus.

The key mechanism that leads to problems is uncontrolled hyperglycemia. Corrigan *et al.* (2009) reported that raised glucose levels in pregnant women with a history of pre-pregnancy diabetes, led to five times greater frequency of cardiac defects than the frequency than occurred in the general population. The cardiac abnormalities most commonly encountered in these cases included transposition of the great arteries, mitral atresia and pulmonary atresia, double outlet of the right ventricle, tetralogy of Fallot and hypertrophic cardiomyopathy.

Specific fetal infections that lead to birth defects

Congenital Rubella virus infection leading to birth defects was first reported by Gregg in 1941. Congenital malformations have also been described in association with cytomegalovirus infection and with intra-uterine toxoplasmosis (Brent, 2004). Congenital cytomegalovirus infection is a major cause of birth defects and hearing loss.

Human immunodeficiency virus (HIV) primarily leads to infection of the fetus during the birthing process and in cases where there is premature rupture of the amniotic membranes and delayed delivery. Treatment of the infant with anti-retroviral medications directly after birth prevents infection. Knapp *et al.* (2012) reported that in a study of 1112 infants born to HIV positive mothers with antiretroviral medication that did not include the known teratogenic efavirenz revealed no association between intra-uterine exposure to antiretrovirals and birth defects. They concluded that, with the exception of efavirenz there was no evidence that antiretrovirals predispose to birth defects.

Fetal Size Duration of Gestation and Onset of Labor

The exact factors that determine natural gestation termination and onset of labor are as yet not definitively characterized. Knowledge about these factors has direct relevance to problems relating to premature birth, low birth-weight and the resulting hazards.

For a number of years the progesterone level has been considered to be the most important factor that determines onset of labor. Progesterone levels are high throughout gestation and apparently serve to reduce uterine muscle contractility. Placental derived progesterone levels fall later in pregnancy and this leads to increased uterine contractility. There is however now evidence that additional factors are important in onset of labor. These factors include progesterone receptor (PGR) isoforms ratios, expression of pro-inflammatory genes, levels of serum proteases and fetal production of surfactant.

Tan *et al.* (2012) reported that the effects of progesterone were modulated by the ratios of expression of the two isoforms of the PGR. PGR isoforms A and B are derived from different promoters and different transcription start sites. They reported that throughout gestation the levels of the PGR B isoforms were higher and this promoted uterine myometrium quiescence. Onset of labor coincided with a rise in the expression of PGR A and reduced PGR B expression. A downstream effect of the increased PGRA expression was increased expression of pro-inflammatory genes, including nuclear factor kappa B (*NFKX*).

Parry *et al.* (2014) carried out proteomic analysis of maternal serum samples obtained between 28 and 32 weeks of gestation. They reported that higher levels of the serum protease serpinB7 were associated with a lower gestational age at delivery.

Hallman (2013) reported that in humans surfactant proteins A, D and C produced by the fetus, influenced the onset of labor. Late in pregnancy surfactants produced in the alveoli of the fetal lung accumulate in the amniotic fluid. Hallman noted that the rise in amniotic surfactant levels signals functional maturity of the fetus and optimal timing of delivery.

It is possible that deleterious mutations or the surfactant encoding genes and reduced expression of normal surfactant may lead to premature onset of labor.

Growth Deficiency and Growth Delays: Genomic and Genetic Factors

Growth deficiency occurs in association with many different chromosome abnormalities including aneuploidies and structural chromosome abnormalities including microdeletions and microduplications. In several syndromes, growth deficiencies occur along with other congenital malformations and/or craniofacial abnormalities, e.g., in Williams–Beuren syndrome (7q11 microdeletion), in Silver–Russell syndrome (duplication of maternally derived 11p11.5).

Growth deficiency is also a feature of a number of inborn errors of metabolism. Specific single gene defects associated with dwarfism and skeletal abnormalities include mutations in fibroblast growth factor receptors (FGFR).

A particularly named syndrome with characteristic phenotypic features that include growth deficiency may be due to defects in any one of a number of different genes. One example is Cornelia de Lange syndrome with features that include growth retardation, limb defects, distinct facial features and developmental delay. Mutation is any one of five different genes which lead to the syndrome. These genes encode protein components of the cohesin complex, SMC1A, SMC3, RAD21 or regulators of the cohesin complex NIPBL and HDAC8. Teresa-Rodrigo *et al.* (2014) reported that 60% of cases of Cornelia de Lange syndrome have NIPBL mutations. Truncating mutations have a severe phenotype, splice mutations lead to a variable phenotype depending on the position of the splice defect. There is evidence that NIPBL may have additional non-cohesin related functions. Zuin *et al.* (2014) published evidence that cohesin binds to a number of different transcription factors and to the promoters of several gene promoters.

Growth hormone deficiency

Mullis (2010) reviewed aspects of growth hormone deficiency. He noted that growth hormone is not a major mediator of skeletal growth and that when growth does not follow the predicted curve a number of factors should be considered, Growth hormone deficiency may however occur due to genetic defects that impact the pituitary development, examples include mutations in the transcription factors PIT1 (POU1F1) and PROP1. Growth hormone deficiency may also arise due to mutations in specific homeobox genes, e.g., *HESX1, SOX2, SOX3, LHX3* and *LHX4*.

Primordial short stature

Primordial short stature is characterized by severe pre- and post-natal growth restriction. This condition is clinically and genetically heterogeneous. Syndromes associated with primordial growth deficiency include Seckel dwarfism, Meier-Gorlin syndrome and 3-M syndrome.

Seckel dwarfism is due to mutations in genes involved in the DNA damage response and DNA repair, e.g., the *ATR* gene that encodes a serine threonine kinase phosphorylates cell cycle checkpoint proteins (O'Driscoll *et al.*, 2007). Kerzendorfer *et al.* (2013) reported that the Meier Gorlin syndrome arises due to mutations in any one of the several genes that play roles in DNA replication. These include genes that encode proteins involved in the recognition of the origin of DNA replication sites e.g., ORC1, ORC4, ORC6 and genes that encode proteins also encodes components of the pre-replication complex CDT1 and CDT6.

In the 3-M syndrome there is severe pre- and post-natal developmental delay with no other developmental abnormalities. Hanson *et al.* (2014) identified defects in three different genes in patients with 3M syndrome, in *CUL7* (cullin 7), *OBSL1* (obscurin like 1) and *CCD8* (coiled–coiled domain 8). Their studies revealed that the proteins encoded by these three genes play important roles at the spliceosome. They demonstrated further that impaired splicing led to reduced expression of the insulin receptor gene (INSR). Reduced expression of INSR likely impairs growth since this receptor binds growth factors.

Russell-Silver dwarfism arises as a result of structural and functional defects in imprinted regions on chromosome 11 and chromosome 7 (discussed further below).

Overgrowth Syndromes and Abnormal Cell Proliferation and Tumor Development During Fetal Life

Sotos syndrome

This syndrome is characterized by overgrowth in height and head circumference over two standard deviations above the mean. There are also unusual facial features including long narrow face, high forehead, downsloping palpebral fissures, prominent narrow jaw and malar flushing.

Patients with this syndrome manifest developmental delay and learning disabilities.

Studies in a patient with Sotos syndrome and a chromosomal translocation with gene disruption, first led to the identification of the NSD1 gene as important in the etiology of the syndrome. NSD1 encodes nuclear receptor binding protein with SET domain and it acts as a transcriptional regulator. Tatton-Brown *et al.* (2012) reported that heterozygous deletions or mutations in the NSD 1 gene occur in 80–90% of patients with Sotos syndrome.

NSD1 acts as a methyltransferase of histone 3 lysine 36 (H3K36). Luscan *et al.* (2014) undertook analysis of eight other H3K36 methyl transferases and 14 H3K27 methyltransferases in patients with Sotos like syndrome and they identified heterozygous mutations in the *SETD2* gene in two patients. They noted that NSD1 and SETD2 encoded enzymes act as histone writers they add methyl groups.

Weaver syndrome

Accelerated bone growth, advanced bone age, large head and skeletal abnormalities including bent joints, particularly in the hands characterize this syndrome. Facial dysmorphology includes wide spaced eyes, low-set ears. Patients also frequently have balance problems and intellectual impairment.

In some cases of Weaver Syndrome the NSD1 gene may be mutated; more frequently the EZH2 gene is mutated. EZH2 encoded a protein that forms part of the polycomb repressive complex PRC2 and it catalyzes the trimethylation of histone 3 lysine 27 H3K27.

DNA methyltransferase *DNMT3A* in overgrowth

Mutations in the DNA methyltransferase enzyme encoded by *DNMT3A* were identified in 13 of 152 cases of overgrowth syndromes studied and these mutations were not present in 1,000 controls (Tatton-Brown *et al.*, 2014). Patients with the *DNMT3A* mutations had overgrowth and unusual facial features including heavy horizontal eyebrows and narrow palpebral fissures.

Beckwith–Wiedemann syndrome

This is an overgrowth syndrome characterized by macrosomia particularly in early childhood. In some cases overgrowth may be asymmetric. Neonates with this syndrome may have visceromegaly and omphalocoele (herniation of the abdominal organs); they may also have large tongues that impair swallowing. Children with this syndrome are at increased risk for Wilms tumor of the kidney, for hepatoblastoma and for rhabdomyomas. This syndrome may result from structural and/or functional defects of genes on chromosome 11p15.

Two imprinted domains occur in the chromosome 11p15 region. One domain includes imprint control region ICR1 with *H19* and *IGF2* gene (insulin like growth factor 2). This region is normally methylated on the paternally derived chromosome and is expressed from the maternally derived genes. The second imprinted domain, the imprinted control region 2, ICR2 has the *KCNQ* and *CDKN1C* genes and is normally methylated on the maternally derived chromosome and is expressed from the paternally derived genes.

Baskin *et al.* (2014) reported that in approximately 9% of patients with Beckwith–Wiedemann syndrome were studied and found to have structural abnormalities of chromosome 11p15. In approximately 25% of patients there was uniparental disomy and both copies of the 11p15 region were derived from the paternal chromosome. They noted that Beckwith–Wiedeman syndrome in some cases resulted from dysregulation at one or more of the imprinted centers on 11p15 or from defects in the genes that regulate imprinting.

Russell–Silver syndrome is characterized by intra-uterine and postnatal growth retardation with normal head growth; asymmetrical growth is sometimes present. Netchine *et al.* (2012) reported that loss of methylation in the ICR1 domain on chromosome 11 occurs in 50% of patients with this syndrome.

More generalized disorders of imprinting

Netchine *et al.* (2012) reported that 25% of patients with Russell–Silver syndrome had loss of methylation in other genomic regions. Eggermann *et al.* (2014) reported that patients with imprinting disorders for example disorders of 11p15 imprinting sometimes show impaired imprinting elsewhere in the genome. These findings indicate the complexity of imprinting and the need to testing for multi-locus methylation defects.

Simpson–Golabi–Behmel overgrowth syndromes

Simpson–Golabi–Behmel syndrome (SGBS) type 1 is a rare prenatal and post-natal overgrowth syndrome and affected individuals have distinctive facial features, congenital malformations, organomegaly and mild to moderate intellectual impairment. This condition maps to the Xq26. Cottereau *et al.* (2014) reported studies in 43 patients with this syndrome who had defects in the glypican 3 gene *GPC3*. In 38% of patients there were exonic deletions, in 24% there were frameshift mutation, in 17% nonsense mutations occurred and in 17% missense mutations were present; in one patient, an exonic duplication occurred.

Glypican 3 is a cell membrane heparan sulfate proteoglycan that plays a key role in growth regulation. Glypicans acts as co-receptors that promote or inhibit signaling in the bone morphogenetic protein (BMP), fibroblast growth factor (FGF), hedgehog (HH) and wingless (WNT) signaling pathways. Dwivedi *et al.* (2013) reported that GPC3 plays roles in endochondral ossification of long bones and intra-membranous ossification of flat bones. GPC3 is also expressed in brain and other organs. In SGBS type 1 *GPC3* mutations typically involve loss of one or more exons, frequently exon 8, leading to production of truncated proteins. Point mutations in the *GPC4* gene also lead to a syndrome with similar phenotypic features.

Simpson–Golabi–Behmel syndrome type 2 is characterized by excessive growth particularly of organs leading to hepatomegaly, nephromegaly, cystic kidneys and skeletal defects with broad thumbs, short fingers, facial abnormalities including short broad nose anteverted nares, high arched palate. Brain malformations include agenesis of the corpus callosum, ventriculomegly and interhemispheric cysts. The features of this syndrome overlap with those of oral facial digital syndrome type 1 OFD1. Both syndromes are due to mutations on the *OFD1* gene that maps to Xp22. Significant variability is likely due to variation in the specific mutations. Female carriers of mutations often have milder symptoms. Bisschoff *et al.* (2012) reviewed clinical findings and genotypes in 30 families classified as oral facial digital syndrome type 1. The *OFD1 gene* has 23 exons. In two cases they determined that the entire *OFD1 gene* was deleted. Splice site and frameshift mutations at points within exons 1 through 16 occurred in most patients with clinical manifestations.

The OFD1 protein is located in the basal body of the cilia and it plays a role in the regulation of synthesis of cilia.

Abnormal Cell Proliferation and Tumor Development in Fetal Life

Kamil *et al.* (2008) reviewed data from a retrospective study of 84 fetuses with tumors. They reported that lymphangiomas and rhabdomyomas were the most common tumors and occurred in 25% and 22% respectively of fetuses studied. Teratomas occurred in 16% and hemangiomas in 14.2% of cases. Tumors that occurred less frequently included glioblastoma of the brain and adrenal neuroblastomas.

Rhabdomyomas

Rhabdomyomas and particularly cardiac rhabdomyomas may arise due to mutation in the tuberous sclerosis genes *TSC1* and *TSC2*. They may also occur in patients with the Beckwith–Wiedemann overgrowth syndrome.

In tuberous sclerosis rhabdomyomas are diagnosed frequently in early infancy; sometimes they are diagnosed prenatally. Cardiac rhabdomyomas may lead to arrhythmias or to obstruction of the cardiac outflow tracts. The rhabdomyomas often regress spontaneously. Loss of heterozygosity is a common finding in tuberous sclerosis tumors (hamartomas) (Green *et al.*, 1994; Sepp *et al.*, 1996). However, the exact frequency of loss of heterozygosity in cardiac rhabdomyomas is not known at the time of writing. Also it is not known if there are differences in the rate of spontaneous regression in cardiac rhabdomyomas dependent upon their molecular characteristics.

Adrenal neuroblastomas

Suprarenal masses detected in fetuses and newborns may be adrenal neuroblastomas. Fisher and Tweedle (2012) reviewed clinical and genetic aspects of neonatal neuroblastomas. Neuroblastomas arise from the sympathetic nerves and are most common in the adrenal gland. Neuroblastomas sometimes metastasize and may lead to compression of the spinal cord.

Mosse *et al.* (2008) and Janoueix-Lerosey *et al.* (2008) have reported that copy number increases in the ALK gene that encodes a receptor tyrosine kinase were frequently observed in neuroblastomas. They also identified mutations in the ALK gene in familial cases of neuroblastomas. Fisher and Tweedle (2012) reported that Alk is mutated in 10% of sporadic neuroblastomas.

Neuroblastomas may occur in association with syndromes due to mutations in the *PHOX2B* gene (encodes a homeobox like transcription factor) or in neurofibromatosis due to *NF1* gene mutations. Neuroblastomas may occur in various positions along the sympathetic chain e.g., in the mediastinum, neck or pelvis.

Maris (2010) reported that in sporadic neuroblastomas, malignant transformation results from interaction of a number of variants of modest effect. There is evidence for enrichment in malignant form of the tumors of specific alleles at the FLJ22536 locus (CASC15 non-protein coding) on 6p22.3 and specific alleles at the *BARD1* gene (encodes a cell growth regulator) at 2p35. In addition, copy number variants of the 1q21 chromosome segment are more common in neuroblastoma cases than controls. Tumors with these segmental copy number variants are also likely to be more aggressive.

Apoptosis Autophagy and Cell Death, in Mammalian Development

Apoptosis and autophagy are related and balanced processes essential in development and throughout life. Key factors in these processes include initiator molecules, effector molecules and signaling pathways and subcellular compartments.

Energy or Amino acid deprivation

⇓

AMPK

TSC1 TSC2 increased   ULK1 increased

mTOR decreased

autophagy stimulated

cell proliferation and growth
decreased

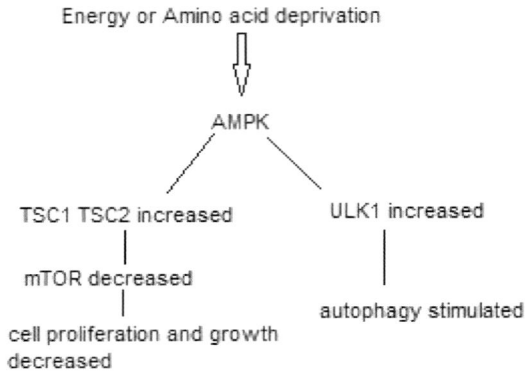

Fig. 5.1: Energy, autophagy.

The metabolic activity and nutrient state of the cell are carefully controlled. In circumstances when nutrient levels are inadequate, autophagy
a catabolic process is initiated. Dunlop and Tee (2013) reviewed
autophagy and key factors. They noted that MTOR and the protein
kinases AMPK and ULK1 (Unc related autophagy related kinase sometimes designated ATK1) play key roles in regulating autophagy. When
cells are energy and nutrient deprived autophagy functions to degrade
macromolecules and redundant or damaged organelles (see Figure 5.1).
Under conditions of energy starvation AMPK transmits positive signaling
to increase ULK1 expression and negative signaling to mTOR occurs.
Negative MTOR signaling results in decreased cell growth and decreased
cell proliferation.

Autophagy is defined as a lysosomal degradation pathway. During
the past two decades at least 10 autophagy related genes (ATG) have been
identified. Mizushima and Levine (2010) reviewed the roles of
autophagy in mammalian development and differentiation. They distinguished three types of autophagy; in macroautophagy cytosolic
components are delivered to lysosomes via autophagosomes.
Microautophagy involves lysosomal uptake through invagination of the
lysosomal membrane. In chaperone mediated autophagy molecules cross
the lysosomal membrane.

Mizushima and Levine (2010) emphasized that the autophagy pathway can drive cellular changes required for differentiation and development.

Autophagy and expression of ATG during pre-implantation of the embryo

Autophagy plays important roles in the early phase of pre-implantation after oocyte fertilization. During this stage maternal mRNA and proteins are degraded and the zygote genome is activated. In mouse, the *Atg5* gene has been shown to play an important role in this process.

During the early neonatal period when placental nutrition is eliminated and external nutrition is not yet adequately established autophagy is necessary to maintain adequate supplies of nutrients.

Particularly interesting aspects of autophagy relate to the process of erythroid differentiation and loss of nucleus and organelles that occur when erythroblasts differentiate to erythrocytes. Mitochondrial clearance from reticulocytes involves a number of different autophagy genes.

Atg7 knockout mice have defects in lymphocyte differentiation. Depletion of Atg7 causes mice to have lymphocytes with greater mitochondrial mass and increased levels of superoxide. Mitochondrial clearance likely plays an important role in lymphocyte differentiation.

Mizushima and Levine noted that in terminally differentiated cells elimination of deformed or dysfunctional mitochondria is important. Damaged mitochondria produced excess reactive oxygen species that are damaging.

Defective autophagy and focal cortical dysplasia

Yasin *et al.* (2013) reported that defective autophagy due to defects in MTOR lead to localized proliferation in the brain. These cortical dysplastic lesions arise during development and may lead to epileptic seizures. Focal cortical dysplasias occur in some patients with tuberous sclerosis. They may however occur in patients with no detectable mutations in the *TSC1* or *TSC2* genes. Yasin *et al.* (2013) determined that balloon cells occur in cortical dysplastic lesions. Within these cells there is lysosomal accumulation. Their data supported the conclusion that MTOR function is not adequately controlled and inhibited in cortical dysplastic lesions.

Apoptosis

Apoptosis occurs as a result of extrinsic signals or as a result of intrinsic signals. Extrinsic signals involve production of TRAIL, a specific ligand in the tumor necrosis factor family and the binding of ligand to FASL, a receptor of the tumor necrosis receptor family. Activation of the receptor leads to activity of the death signaling initiator complex and activation of caspases See Figure 5.3.

Intrinsic signals include DNA damage. This leads to activation of the pro-apoptotic members of the BCL family of proteins BAX and BAK. This activation leads to increased permeabiltiy of the mitochondrial membrane and release of cytochrome C. The anti-apoptotic BCL family proteins BCL, BCL2 and BCLX may inhibit the process. The cytosol protein APAF1 (APAF1 (apoptotic peptidase activating factor) senses released cytochrome C. It undergoes oligomerization and triggers expression of Caspase 9, the executioner caspase. Riedl and Salveson (2007) reported that in the presence of cytochrome C, APAF protein oligomerizes to form a wheel like signaling platform, the apoptosome (see Figure 5.2).

There is evidence that cross talk occurs between autophagy and apoptosis. This is likely mediated by the autophagy related protein beclin that binds to the apoptosis related BCL proteins and to the ATG proteins

There is evidence that when expression of genes involved in apoptosis pathways is impaired brain malformation result. Programmed cell death is

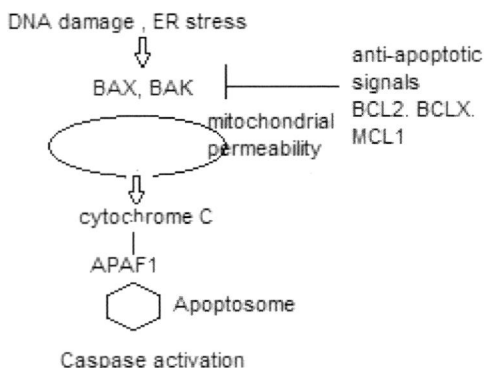

Fig. 5.2:   Apoptosis signaling.

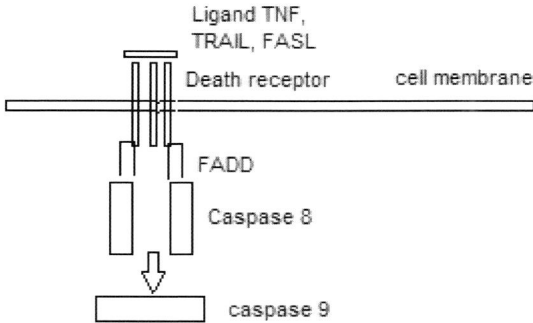

Fig. 5.3: Cell death signaling.

important in the development of the central nervous system and programmed cell death impacts maturation of the cerebral cortex, cerebellum, thalamus, brainstem and spinal cord (Lossi and Merighi, 2003).

The Hippo Signaling Pathway

The Hippo pathway fine-tunes cell proliferation and cell death and plays key roles in the determination of organ size. Varelas and Wrana (2011) reported that through intra-cellular signaling the Hippo pathway plays a role in the determination of cell proliferation, cell death and co-ordinates organ size. The Hippo pathway includes a cascade of serine threonine kinases that activate transcriptional co-activators. These then play roles in activating transcriptional co-activators that then bind to DNA. The Hippo pathway is also connected to the TGF beta and WNT signaling pathways.

CHAPTER 6

METABOLISM AND ORGANELLES

Inborn Errors of Metabolism That are Present in Newborns

A critical period of development includes the transition from intrauterine to post-natal life. In post-natal life the tissue of the infant will become totally responsible for performance of essential metabolic functions. Following a normal pregnancy and birth within a few days of life the condition a child with an inborn error of metabolism may suddenly deteriorate and an acute neurological crisis may result.

Saudubray *et al.* (2002) reported that approximately 100 different metabolic errors could first be present in the newborn period. The symptoms of these conditions often overlap with other deleterious conditions in the newborn including poor feeding, prolonged or severe jaundice and sepsis. Since many of the inborn errors are inherited as recessive conditions, there may be no family history of a disorder and no history of previous infants with similar manifestations.

Clinical manifestations of metabolic diseases in the newborn period

Saudubray *et al.* (2002) distinguished three different categories of disorders and manifestations during this period, intoxication disorders, energy deficiency disorders, and disorders of complex molecules.

Intoxication disorders

Disorders in this category lead to accumulation of toxic metabolites. Examples are abnormalities of metabolism that lead to accumulation of abnormal levels or abnormal types of organic acids e.g., methylmalonic acidemia, lactic acidemia. Examples include inborn errors of amino acid metabolism, e.g., defects in metabolism of branched chain amino acids leading to maple syrup urine disease. Urea cycle disorders may lead to toxic accumulations of ammonia. Infants with these disorders frequently present with vomiting and then with lethargy and coma. Lactic acidemia may result from defects in biotin metabolism and holocarboxylase synthase function.

Disorders of energy metabolism

In this category Saudubray *et al.* (2002) included disorders where energy production or utilization was defective. Defective metabolism of glucose and increased levels of lactic acid occur when pyruvate carboxylase or pyruvate dehydrogenase complexes are deficient. In mitochondrial respiratory chain disorders impaired oxidative phosphorylation leads to impaired energy production. Defective release of glucose from glycogen due to abnormalities of glycogen metabolism, lead to energy depletion.

Fatty acid oxidation defects may also lead to energy depletion. Fatty acids are transported to mitochondria for beta-oxidation and this transport requires carnitine. Some fatty acid oxidation disorders can be treated with carnitine supplementation and increased carbohydrate intake. Energy depletion disorders often present with low muscle tone, poor cardiac function including arrhythmias and heart failure.

Important causes of acidosis include abnormalities of the pyruvate dehydrogenase complex and disorders of the pyruvate carboxylase complex.

Pyruvate dehydrogenase complex catalyzes conversion of pyruvate to acetyl Coenzyme A. The complex components and chromosome locations of encoding genes are as follows: E1 pyruvate dehydrogenase containing two subunits PDHA1 (Xp22.1) and PDHB (3p21.1), E2 dihydrolipoyl transacetylase (DLAT) (11q22.1), and E3 dihydrolipoyl dehydrogenase (DLD) (7q31-q32). PDHX on 11p13 encodes an important E3 binding protein.

Pyruvate carboxylase is encoded by a single locus on chromosome 11q13.4. The active enzyme is a homotetramer, requires biotin and adenosine triphosphate (ATP), and converts pyruvate to oxaloacetate.

Maple Syrup urine disease

In this disease there is impaired catabolism of the branched-chain aminoacids leucine, isoleucine, and valine. The site of the defect is at the step where the derived oxo-acid is cleaved, i.e., R-CO-COOH to

R-CO-acetylcoenzyme A. This cleavage requires branched-chain keto-acid dehydrogenase and thiamine pyrophosphate, and nicotinamide adenine dinucleotide (NAD). The branched-chain keto dehydrogenase enzyme is composed of three different subunits. The subunits and the chromosomal locations of the genes that encode them are as follows: Branched-chain keto-acid dehydrogenase subunit A (BCKDHA) (19q13.1), subunit B (BCKDHB) (6q14.1), dihydrolipoyl dehydrogenase (DLD) (7q31-q32), and dihydrolipoamide branched-chain transacylase (DBT) (1p31).

Methylmalonic academia

Patients with the organic acidemia, methylmalonic acidemia frequently have defect in the enzymes necessary for processing Cobalamin (Vitamin B12) to cofactors necessary for enzymes involved in transfer of methyl residues. Froese and Gravel (2010) reported that defects in generation of these cofactors might lead to devastating illness in new-borns or in early childhood. Ingested Vitamin B12 binds to haptocorrin released in the upper gastrointestinal tract. Cobalamin then binds to intrinsic factor. It then binds to a Cobalamin receptors (Cubulins) on the apical surface of intestinal cells in the ileum and is taken up into those cells. In the circulation, it binds to transcobalamin 2 (TCN2) and is taken up into cells through receptor endocytosis and enters lysosomes. There it undergoes digestion to free cobalamin that then enters the cyto-plasm and undergoes transformation to produce the cofactors methylcobalamin and adenosylcobalamin. In some cases cobalamin remains trapped in the lysosomes due to defects in a lysosomal mem-brane protein LMBD1. The enzyme methylmalonic aciduria and homocystinuria C (MMAHCC) (1q34.1) catalyzes an early step in the modification of Cobalamin. The protein methylmalonic aciduria and homocystinuria type D (MMADHC) (2q23.2) may act as a chaperone. The enzymes methylmalonic aciduria type A(MMAA) (4q31.21) and methylmalonic aciduria type B (MMAB) (12q24) are responsible for transformation to adenosylcobalamin. Adenosylcobalamin is required as cofactor for the enzyme methylmalonyl-CoAmutase (MUT) to convert L-methylmalonyl-CoA to succinyl-CoA. The enzyme MTRR (methyltransferase reductase) (5p15.31) generates methionine from homocysteine with cofactors methylcobalamin and tetrahydrofolate. Defects in any of these enzymes may lead to methylmalonic acidemia with or without homocysteinemia.

Urea cycle disorders

In the neonatal period urea cycle defects may present with neurologic deterioration including seizures and hypotonia. They may also present with hypoglycemia and impaired liver functions. These disorders lead to significant elevations of blood ammonia levels. Fluctuations in ammonia levels in response to alterations in protein intake also provide useful information. The specific site in the pathway of the enzyme block can often be determined by establishing relative levels of arginine, citrulline and ornithine, arginine, arginosuccinic acid, and orotic acid.

Multiple carboxylase deficiency and biotinidase deficiency

Multiple carboxylase deficiency due to defective function of holocarboxylase synthetase often presents with overwhelming illness very early in post-natal life. This includes profound acidosis, neurologic deterioration, and significant dermatological disorders with significant erythema and peeling of skin. Holocarboxylase synthetase requires biotin for its formation and function. Biotin is the required cofactor for conversion of apocarboxylase to holocarboxylase. Biochemical manifestations of this deficiency include elevated levels of lactic acid in plasma and urine and also increased urinary excretion of specific organic acids, including 3-hydroxybutyric acid and 3-methylcrotonylic acid.

The enzyme biotinidase (see Figure 6.1) is required for release of free biotin from its bound form, biocytin. Biotinase deficiency may be present in infants or may occur later in life. Symptoms may be neurologic (seizures, spasticity, tremors, deafness) or dermatologic in nature (see Table 6.1). Unusual levels and types of organic acids may be present in the urine.

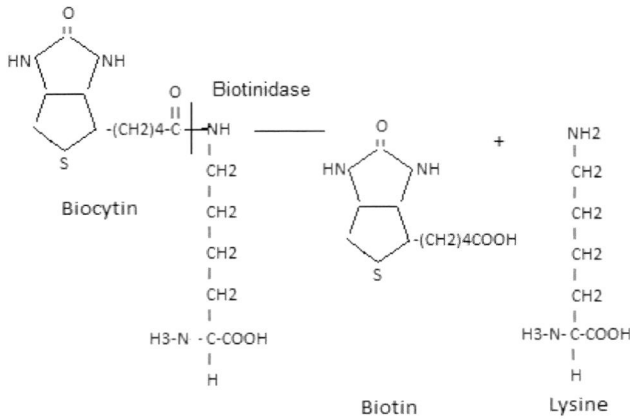

Fig. 6.1:   Biotinidase.

Table 6.1:   Neonatal Seizure Diseases Causes and Diagnostic Testing.

Disorders	Testing/Causes
Pyridoxine dependent seizures	Pipecolic acid level, Aldehyde dehydrogenase(ALDH7A1) gene
Pyridoxal phosphate dependent seizures	Pyridoxamine-5'-phosphate oxidase (PNPO) gene
Serine biosynthesis defect	3-phosphoglycerate dehydrogenase (PHGDH) genes
Glucose Transporter Type (GLUT1) deficiency	Fluorodeoxyglucose (FDG PET) scan; SLC2A1 gene
Non-ketotic hyperglycinemia	Glycine cleavage complex activity
Sulfite oxidase SUOX molydenium cofactor deficient	urine sulfites, sulfocysteine, homocysteine
Congenital neuronal ceroid lipofuscinoses	CTSD gene (Cathepsin D)
Dihydropyrimidine dehydrogenase deficiency	Gene testing DPYD
Creatine deficiency syndromes	Serum and urine creatine; specific genes
Biotinidase deficiency	Biotinidase assay, BTD gene

Disorders of Complex Molecules

Lysosomal and peroxisomal disorders

This category includes lysosomal and peroxisomal disorders and congenital defects of glycosylation. Many forms of these disorders present later than the newborn period. However, several forms of peroxisomal disorders, such as Zellweger syndrome, present at birth with neurological dysfunction and unusual facial features. (Further details on lysosomes and peroxisomes are presented in the Organelle section.)

Lysosomal storage diseases occur due to deficiency in the breakdown of complex molecules e.g., glycosphingolipids usually present later in life. However, the infantile form of Niemann–Pick disease, due to sphingomyelinase deficiency, may present with neonatal edema and even before birth with hydrops fetalis (abnormal fluid accumulation in body cavities and swelling). The spleen may be enlarged at birth and abnormal storage of sphingomyelin may be present in liver, kidney and placenta (Nyhan *et al.*, 2005).

The acute infantile form of Gaucher disease due to glucocerebrosidase deficiency may present within the first three months of life, but earlier occurrence with hydrops fetalis has also been described for this disease (Nyhan *et al.*, 2005).

Severe infantile epilepsy syndrome may occur in young infants due to defects in the biosynthesis of gangliosides e.g., GM3 synthase deficiency, (Platt 2014). GM3 is monosialodihexosylganglioside. GM3 synthase activity is necessary in cell differentiation and proliferation and in signal transduction. GM3 synthase uses lactosyl-ceramide for synthesis of gangliosides.

Congenital disorders of glycosylation (CDG)

Patients with congenital disorders of glycosylation may present with intra-uterine growth retardation or with hypotonia, or seizures. Glycosylation of protein involves the addition of activated nucleotide sugars, most frequently mannose, to specific protein residues. Specific pathways include N-linked glycosylation where nucleotide sugars bind to asparagine and O-linked glycosylation, where nucleotide sugars bind to serine or threonine residues. In congenital disorders of glycosylation there are frequently defects in the synthesis of nucleotide sugars. Key steps in the generation of activated sugars are generation of mannose-6-phosphate, and isomerization to mannose1-phosphate and formation of the nucleotide GDP-mannose. Deficiency in the enzymes phosphomannomutase PMM1 and PMM2 are frequent causes of congenital disorders of glycosylation. CDG Type 1a, phosphomannomutase type 2 deficiency, may present with fetal hydrops and may occur in the newborn period with neurological symptoms including hypotonia and/or with cardiomyopathy. This disorder results from deficiency of the enzyme phosphomannomutase 2 and deficiency leads to defective glycosylation of proteins (van de Kamp *et al.*, 2007).

Defects in glycosyltransferases (e.g., ALG6) alpha-1, 3-glucosyltransferases that attach nucleotide sugars to asparagine are also responsible for some cases of congenital disorders of glycosylation and seizures.

Seizures in the Newborn Period

It is important to emphasize that a number of inborn errors of metabolism may present with seizures in the newborn period in the absence of biochemical abnormalities such as acidosis, ketosis, or hyperammonemia. Ficicioglu and Bearden (2011) reviewed causes of newborn seizures. These investigators emphasized the importance of accurate diagnosis since beneficial and specific treatments are available for a number of these disorders, though sadly not for all. Examples of treatable newborn seizure disorders include pyridoxine and pyridoxine phosphate dependent seizures, folinic acid responsive seizures, glucose transporter type 1 (GLUT1) deficiency, creatine deficiency syndromes and serine biosynthesis defects (See Table 6.1).

Pyridoxine responsive seizures

Defective function of the enzyme ALDH7A1, also known as antiquitin, leads to seizures that may have neonatal onset and to encephalopathy. These seizures can often be controlled by large doses of pyridoxine. ALDH7A1 defects can also lead to seizures that are responsive to folinic acid. ALDH7A1 functions in the lysine degradation pathway. In this pathway lysine is converted to saccharopine and then to alpha-aminosemialdehyde, ALDH7A1 functions as an alpha-aminosemialdehyde dehydrogenase to generate alpha-aminoadipic acid. Lysine degradation in a different pathway leads to Pipecolic acid formation. When ALDH7A1 is deficient, excess quantities of Pipecolic acid are formed.

In some infants neonatal seizures that do not respond to pyridoxine treatment do respond to treatment with pyridoxine-5′-phophate. In some of these infants defects in the function of the enzyme PNPO pyridoxamine-5′-phosphate oxidase were identified. This enzyme catalyzes reactions in the rate limiting steps leading to formation of the biologically active form of pyridoxine (vitaminB6), (Stockler *et al.*, 2011).

Sulfite oxidase deficiency

Methionine and cysteine metabolism generate sulfites, with sulfite oxidase deficiency these cannot be efficiently metabolized, leading to accumulation of sulfites. Ficicioglu and Bearden reported that the accumulation of sulfites particularly decreases NAD activity and impairs synthesis of adenosine triphosphate.

A molybdenum binding cofactor is important for function of a number of different oxidases, including sulfite oxidase, and xanthine oxidase. Synthesis of this cofactor includes four steps and interactions with pterin molecules. Deficiency of sulfite oxidase leads to seizures and may lead to multicystic lesions in the cerebral cortex (Higuchi *et al.*, 2014).

Serine biogenesis defects

Serine is a precursor of the neurotransmitter glycine and a precursor of phospholipids that form myelin Ficicioglu and Bearden (2011). Defects in serine synthesis and low levels of serine in cerebrospinal fluid and brain predispose to seizures that usually occur first in the newborn period. Most cases of these disorders result from defects in the function of the enzyme 3-phosphoglycerate dehydrogenase that concerts 3-phosphoglycerate to 3-phosphopyruvate. Defects in the two subsequent steps in serine biosynthesis are less common. These steps include conversion of 3-phosphopyruvate to 3-phosphoserine, the activity of phosphoserine aminotransferase (PSAT), and conversion of 3-phosphoserine to L-serine through phosphoserine phosphatase (PSP) activity.

Non-ketotic hyperglycinemia

This disorder results from defective activity of cleavage actions and defective synthesis of neurotransmitters. Glycine cleavage requires four enzyme components and in addition uses cofactors tetrahydrofolate (THF) and NAD. The four enzymes components are P-protein, pyridoxal phosphate containing protein, T-protein tetrahydrofolate-binding protein, H-protein lipoic acid containing protein, and L-protein lipoamide dehydrogenases. The H-protein (GCSH) shuttles the methylamine group of glycine from the P-protein (GLDC) to the T-protein (GCST). The neonatal form of this disorder leads to seizures and encephalopathy manifesting usually by 48 hours after birth.

Neuronal ceroid lipofuscinoses

Defects in any one of the products of at least 10 different genes lead to storage of an abnormal autofluorescent material lipofucsin. Storage occurs particularly in brain. The form of ceroid lipofuscinosis (CLN10) due to defects in the Cathepsin D gene leads to neonatal seizures. This disorder may also lead to microcephaly and intra-uterine seizures. Cathepsin D enzyme cleaves peptide bonds.

Cerebral creatine deficiency syndromes

In brain and muscle the creatine/creatine phosphate system plays a key role in maintaining energy. Comeaux *et al.* (2013) described three cerebral creatine deficiency syndromes (CCDS). These syndromes are characterized by absence or reduced levels of creatine in brain as assessed by proton magnetic resonance spectroscopy. In two forms of the disorder creatine synthesis is impacted and in the third form creatine transport is defective.

In the first step of creatine synthesis arginine and glycine are transformed through the activity of the enzyme l-arginine: glycine amidinotransferase (AGAT) to form guanidine-acetate and to release ornithine.

In the second reaction of creatine synthesis in the presence of S-adenosylmethionine as methyl donor guanidine-acetate is transformed to creatine through activity of the enzyme guanidinoacetate methyltransferase (GAMT) and S-adenosyl-homocysteine is generated.

Synthesis of creatine occurs in the liver and kidney. Creatine is then transferred across cell membranes through activity of the solute transporter SLC6A8 encoded by a gene on chromosome Xq28. Within brain and muscle creatine kinase transforms creatine-to-creatine phosphate. Comeaux *et al.* (2013) noted that there is a broad range of clinical severity in these disorders. The diagnosis of impaired activity of AGAT or GAMT enzymes can be facilitated through assessment of plasma levels of guanidine-acetate and creatine.

Defective creatine transporter levels lead to elevated ratios of creatine to creatinine in males. In females this assay is less reliable. Molecular deoxyribonucleic acid (DNA) gene studies are frequently carried out to confirm diagnosis.

Inborn Errors of Metabolism Associated with Severe Jaundice or Liver Failure

Bile acid synthesis disorders may lead to severe jaundice in newborns, (see section on bile synthesis in chapter on Liver development). Severe jaundice and compromised liver function may also occur in specific forms of galactosemia, in fructose intolerance and in tyrosinemia.

Galactosemia results from mutations in the galactose-1-phosphate uridyl transferase. However, many variants occur in this gene and these lead to slight alterations in function that are not clinically significant. Clinically important defects in this enzyme are associated with increased levels of galactose-1-phosphate and the patient is at risk for severe jaundice compromised liver function and sepsis (Berry, 2012).

Organelles

Mitochondria

Human mitochondrial DNA is maternally inherited; mitochondria derived from the oocyte are the only mitochondria present in the early embryo.

It is important to note that each oocyte contains many mitochondria and that the DNA sequence present in each mitochondrion is not necessarily identical. Replication of mitochondria does not begin until after implantation. Van Blerkom (2008) reported that during the pre-implantation period mitochondria in the early embryo undergo structural transformation, they elongate and develop extensive arrays of cristae. Subsequent mitochondrial biogenesis and mitochondrial roles in energy production are essential for embryonic growth and development. Different cells and tissues frequently differ in the exact number of mitochondria present, and mitochondria with different DNA sequences may be present in different proportions in different cells. This condition is referred to as heteroplasmy.

The mitochondrial DNA encodes 13 proteins, 22 transfer ribonucleic acid(RNAs) and 2 ribosomal RNAs. As new mitochondria develop in the embryo they are populated not only by the products encoded in the mitochondrial DNA but also by nuclear encoded gene products that are targeted to the mitochondria. These include more than 1,100 components (Vafai and Mootha, 2012).

Mitochondrial DNA replication

It is important to note that nuclear encoded gene products are also required for synthesis of new mitochondria (mitochondrial biogenesis) and for synthesis of mitochondrial DNA. At least 20 different nuclear encoded gene products constitute the mitochondrial DNA replisome. All of these products need to be transported into the mitochondria. Deoxynucleotides required for DNA synthesis are imported into mitochondria from the cytoplasm. Deoxynucleotides are in part derived by synthesis of deoxyribonucleotides from cytosolic ribonucleotides through activity of the ribonucleotide reductase enzymes (RRM1 and RRM2). Deficiency of specific products required for DNA synthesis and mitochondrial replication lead to mitochondrial depletion syndromes (discussed later in this chapter).

Mitochondrial biogenesis

The numbers of mitochondria present in each cell are impacted by organelle biogenesis, by fission and fusion of mitochondria and by mitophagy.

Scarpulla (2011) reported that the Peroxisome proliferator-activated receptor gamma coactivator also known as PGC (sometimes designated as PPARGC) family of transcription co-activators plays important roles in regulating the rate of mitochondrial biogenesis. There is evidence that energy sensors, including the Adenosine monoposphate(AMP) kinase and sirtuins, mediate expression of the PGC1 family members. Sirtuins are NAD dependent enzymes. The nuclear sirtuins SIRT1, SIRT6 and SIRT7 regulate transcription factors and metabolic pathways. SIRT3, SIRT4 and SIRT5 function in mitochondria and regulate activity of mitochondrial enzymes (Pirinen *et al.*, 2012). AMP kinase is a crucial cellular energy sensor that promotes ATP production when energy levels are falling. It also impacts cell cycle activity and neuronal membrane excitability.

Mitochondrial fission and fusion

Fission involves processes of splitting off to form a new mitochondrion from an existing mitochondrion. Fusion and fission are also critical processes in determining mitochondrial numbers. Fusion between mitochondria also to some extent reduces differences between mitochondria. Fusion also plays a role in maintaining a healthy overall state of mitochondria (Vafai and Mootha, 2012). Mitofusins are proteins encoded by the MFN1 and MFN2 genes that are involved in fusion of the outer mitochondria membranes. Heterozygous mutations in the MFN2 gene lead to a peripheral axonal neuropathy CMT2 (Charcot Marie Neuropathy type 2A).

Fusion of the inner mitochondrial membranes requires optic atrophy (*OPA1*) gene. This gene is defective in a specific form of optic atrophy in humans. The product of this gene helps to determine the mitochondrial morphology that is essential for normal function. OPA1 mutations are associated with mitochondrial fragmentation and disruption of mitochondrial cristae. The normal *OPA1* gene generates different mRNA transcripts and different protein isoforms as a result of differential splicing. Defects in mitochondrial fusion are also associated with deficiency in mitochondria DNA synthesis.

Chan (2012) reported that balance between rates of mitochondrial fusion and fission is important for healthy mitochondria function. One protein that plays an important role in fission is dynamin encoded by the dynamin related protein (*DRP1*) gene.

Mitophagy

Mitophagy plays an important role in removing dysfunctional mitochondria. Specific proteins involved in this process include Parkin PARK2 (a form of ubiquitin ligase) and PINK1 (Pten induced kinase) Narendra *et al.*, 2012. PINK1 recruits PARK2 to the outer membrane of damaged mitochondria and PARK2 then ubiquitinates outer membrane proteins and these are then transported to autophagosomes for digestion.

Mitochondrial DNA transcription

Transcription of mitochondrial DNA is indicative of the bacterial origin of mitochondria (Vafai and Mootha, 2012). The 16-kilobase circular DNA in mitochondria is present as a heavy chain and a light chain that are transcribed as two continuous polycistrons one heavy chain polycistron and one light chain polycistron. Cleavage of the heavy chain polycistronic transcript gives rise to 12 mRNAs for enzyme proteins, 14 transfer RNAs and 2 ribosomal RNAs. The light chain polycistronic transcript is cleaved to give rise to 8 transfer RNAs and one protein encoding mRNA for the complex 1 component ND6. Specific mitochondrial RNA processing proteins have been identified (Wolf and Mootha, 2014).

Mitochondrial respiratory complexes and oxidative phosphorylation

Nuclear genes encode products that facilitate transport of proteins and enzymes into mitochondria and products that play roles in assembly of multiprotein complexes in mitochondria. Nuclear encoding components are the predominant proteins in the mitochondrial respiratory complexes.

The key components for energy production by mitochondria include the respiratory complexes. Complexes I and II transfer electrons from the metabolism derived coenzymes Nicotinamide adenine dinucleotide (NADH) and Flavin adenine dinucleotide(FADH) to coenzyme Q. Complex III then transfers electrons from reduced Coenzyme Q to cytochrome C. Electrons are then transferred to complex IV and used to

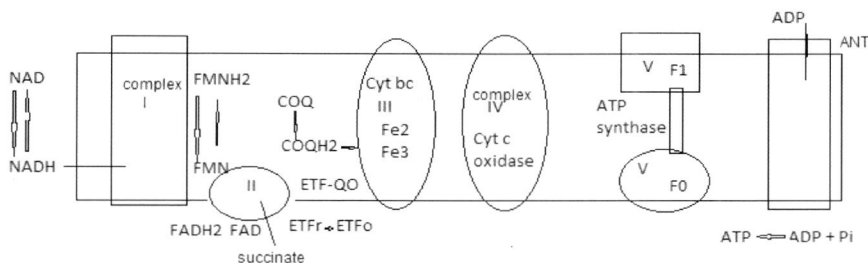

Fig. 6.2:   Mito Elec Trans (Mitochondrial Electron Transport).

reduce molecular oxygen (O2). The transfer of electrons from complexes I, III, IV is coupled to the formation of a proton gradient in the mitochondrial inter-membrane space. Complex V then utilizes this proton motive force for ATP synthesis (see Figure 6.2).

The majority of the components of the mitochondrial respiratory complexes are nuclear encoded; generation of these complexes therefore requires cross talk between nuclear and mitochondrial genomes. Regulatory factors involved in this cross-talk are not yet fully characterized. Research into factors involved in the assembly of the respiratory complexes is ongoing, however a number of assembly factors have been identified.

Complex 1 NADH ubiquinone oxidoreductase is composed of 45 different subunits. There are 14 core functional subunits, 7 are nuclear encoded, 7 are mitochondrial encoded. The other components are key assembly factors and stabilization factors. Defects in core subunits or in assembly factors may lead to disease including lactic acidosis or cardiomyopathy.

Analysis of the nuclear gene sequences that encode proteins that function in mitochondria and studies on patients have facilitated development of a growing body of data on the subunit composition of the all four of the respiratory complexes and has led to the identification of many of the accessory factors required for assembly and stabilization of complexes. It has also become clear that many cases of mitochondrial disease, including those that present in the newborn period with acidosis, encephalopathy and cardiomyopathy are due to mutations that impact functions of assembly factors.

Other mitochondrial functions

Many other functions are carried within mitochondria. These include lipid, ketone and amino acid metabolism, roles in steroid hormone synthesis, and synthesis of iron (Fe) containing proteins; mitochondria also play an important role in controlling reactive oxygen species generation and in calcium homeostasis. There are likely many other functions carried out in mitochondria that remain to be discovered. Elucidation of the extent of nuclear encoding of mitochondrial components was only achieved in 2008 when an accurate inventory of the mitochondrial proteome and the MitoCarta database were compiled (Pagliarini *et al.*, 2008).

It is also likely that additional functions will be determined for currently known components of mitochondria (Vafai and Mootha, 2012). It is important to note that although all cells and tissues have mitochondria, there are tissue and cell specific differences in mitochondrial number, mitochondrial protein composition and Oxidative phosphorylation(OXPHOS) capacity. Pagliarini *et al.* (2008) reported that on average organs share 75% of their mitochondrial components and that tissue heterogeneity impacts function.

There is evidence for mitochondrial functional heterogeneity in different tissues and organs. Vafai and Mootha (2012) noted that in skeletal muscle mitochondrial oxidation of fatty acids predominates, while in brain oxidation of ketones predominates, and in the adrenal gland steroid synthesis predominates.

Mitochondrial transporters

There is now information on the key role of transporters that carry solutes into mitochondria. The mitochondrial transporters (SLC25) family includes 53 different proteins that shuttle solutes including aminoacids, nucleotides, dinucleotides, carnitine, acylcarnitine and carboxylates (e.g., alpha-ketoglutarate, oxaloacetic acid, citrate succinate) across mitochondrial membranes. At least 14 different inherited diseases are known to result from defects in specific SLC25 proteins (Palmieri, 2014). Neonatal epileptic encephalopathy results from defective function of SLC25A22 (mitochondrial glutamate carrier).

Mitochondrial malfunction syndromes in infants

In some cases function of multiple mitochondrial respiratory complexes may be impaired. These can result from defects in the pyruvate dehydrogenase complex and impaired generation of NADH or from impaired biosynthesis of iron sulfur complexes. Infants with these conditions may present with neonatal lactic acidosis, hypotonia, cardiomyopathy, and seizures.

Leigh syndrome is defined as a progressive neurodegenerative disorder in infant with seizures and increased levels of lactate particularly in the cerebrospinal fluid. It is commonly due to impaired oxidative phosphorylation.

Mitochondrial depletion syndromes

In these conditions the numbers of mitochondria are reduced. Manifestations may be present early in life and they include cerebral, hepatic, and intestinal malfunction (Nogueira *et al.*, 2014).

Key factors in the etiology of mitochondrial depletions syndrome include defects in proteins involved in synthesis of mitochondrial DNA e.g., defects in polymerase gamma (POLG or Twinkle). In addition defects in purine salvage or pyrimidine synthesis may lead to mitochondrial DNA synthesis defects. Deficiency of the enzyme ribonucleotide reductase (RRM2B) that is required for synthesis of deoxynucleotides from cytosolic ribonucleotides leads to mitochondrial deficiency syndromes.

Lysosomes and Peroxisomes

Lysosomes serve as the primary catabolic compartment in cells and degrade components that have been taken up from the interior of the cell and beyond the cell membrane. Newer studies have revealed that lysosomes have important roles in energy metabolism and in mediating cellular response to changing levels of nutrients. Settembre *et al.* (2013) identified a transcription factor TFEB (transcription factor EB) that acts as a master regulator of lysosomal biogenesis in response to environmental conditions.

There is evidence that lysosomal function is important from early embryonic life onward. Tsukamoto *et al.* (2013) carried out studies in mice and reported that in the early embryo inhibition of expression of genes that encode lysosomal membrane proteins, LAMP1 and LAMP2, led to developmental arrest. They noted that pharmacological inhibition of lysosomal function also led to developmental retardation.

Lysosomal functions

Keys to lysosomal function are lysosomal membrane proteins and the approximately 50 lysosomal hydrolases. Saftig and Klumperman (2009) reviewed lysosomal proteins and functions.

Lysosomal hydrolases are released from the trans-Golgi network and are targeted to lysosomes through the mannose6-phosphate receptor (M6R). However, there is evidence from patient studies that in the absence of efficient M6R function hydrolases do reach lysosomes but specific hydrolases are reduced in quantity. Other receptor systems are involved in lysosomal enzyme uptake. These include the vacuolar protein-sorting(VPS) domain containing sortilin proteins encoded by genes *SORCS1, SORCS2*. Sortilins are particularly important in uptake of sphingolmyelinase and sphingolipid activator protein.

Saftig and Klumperman (2009) noted that beta-glucocerebrosidase (the enzyme that is deficient in Gaucher disease) is not taken up using the M6R system. Its transfer is dependent on LIMP2, a glycoprotein expressed in membranes of lysosomes and endosomes. Deficiency of LIMP2 leads to myoclonic epilepsy. Sortilins also transfer cathepsins (cysteine proteinases) to lysosomes.

A number of different storage diseases result from mutations that impact lysosomal hydrolases or the proteins involved in transfer across lysosomal membranes. Saftig and Klumperman (2009) emphasize that there is growing realization that a number of different diseases are due to defects in proteins involved in the transfer across lysosomal membranes. These include defects in: cation and H^+ transfer, organic compound transfer, transfer of complex molecules including sialic acid, cholesterol and lipids.

Cellular pathways in lysosomal biogenesis and functions

Saftig and Klumperman (2009) emphasized lysosomes receive material through fusion with endosomes. Material taken up at the cell membrane by phagocytosis passes through a series of different forms of phago-somes and endosomes and ultimately to the lysosome. Macroau-tophagy involves digestion of large complexes and organelles. Endosomes also take up material from the cellular interior. There are also lysosomal exocytosis pathways that deliver proteins to the outside of the cell.

Cholesterol homeostasis involved lysosomes and the NPC1 protein (Niemann–Pick C1) that impacts cholesterol import and export. Mucolipin-1(MCOLN1) is an ion channel with transmembrane domains. MCOLN1 is deficient in the autosomal recessive disease mucolipidosis 1. Heparan alpha glucosaminide N-acetyltransferase (HGSNAT) is deficient in the lysosomal storage disease, Sanfilippo syndrome Type 3. HGSNAT is an integral membrane protein with transmembrane domains that degrades heparan sulfate. It exists as a large precursor from which smaller proteins are generated. However, both the precursor protein and the smaller proteins have enzyme activity.

Peroxisomes

Single membranes surround peroxisomes. Peroxisomes contain about 100 enzymes that are involved in beta-oxidation of fatty acids, synthesis of lipids and bile acid metabolism.

Studies in specific genetic disorders e.g., Zellweger syndrome and related disorders have revealed that specific genes encode proteins that are required for peroxisomal membrane assembly and additional genes encode proteins and enzymes active in peroxisomes. Specific clinical syndromes have been identified that result from deficiency of single peroxisomal enzymes or proteins. In addition there are disorders where multiple per-oxisomal enzymes are missing and these are due to peroxisomal biogenesis defects, (Wanders, 2013). Those 13 genes that encode products involved in peroxisomal biogenesis have been cloned.

Cilia

Cilia are present throughout development. Primary cilia are likely present on all cells and are the location of receptors. Motile cilia occur in clusters and through sweeping movements they control fluid movements. In the brain cilia are involved in cell migration, axon guidance, and regional patterning. Motile cilia impact the movement of cerebro–spinal fluid.

Cilia are composed of more than 1,000 different protein and primary cilia constitute a signaling platform for the cells. Primary cilia are microtubule-based organelles that occur on cell surfaces. Hildebrandt *et al.* (2011) emphasized that receptors on primary cilia receive a broad range of extra-cellular signals and that these signals are transmitted to the interior of the cells. The lipid bilayer membranes of cilia harbor tyrosine kinase receptors, G-protein coupled receptors, ion channels, Notch receptors, ion channels and transporter proteins.

Cilia are comprised of a microtubule-containing core, the axoneme. Motile cilia are 9 + 2 cilia; in these cilia the axoneme is composed of 9 doublet microtubules and 2 central microtubules. In motile cilia the microtubule doublets are connected by dyneins and radial spokes connect the outer doublets with the inner doublets. In primary cilia the central doublet is absent and these cilia are sometimes referred to as 9 + 0 cilia.

In both these types of cilia a lipid bilayer membrane surrounds the axoneme. The axoneme is anchored to the basal body. The axoneme is formed from the basal body of a mother centriole. A transition zone and transition fibers surround the region where the cilia borders on the membrane. The transition zone fibers filter molecules that pass from the cell to the cilia. In addition continual growth of the cilium requires transport of protein.

Christensen *et al.* (2012) reported that within the axoneme macromolecular particles and membrane proteins are transported in an intra-flagellar transport process along the microtubules (see Figure 6.3). Anterograde transport utilizes kinesin 2 composed of KIF3a and KIF3b and a non-motor subunit. Retrograde transport utilizes dynein 1B. The transition between anterograde and retrograde transport occurs at the axoneme tip complex.

Tubulins encoded by *TUB1A* and *TUB1B* occur in the microtubules. Nexins stabilize axonemes in cilia. Mutations in TUB1A have been

Cilium

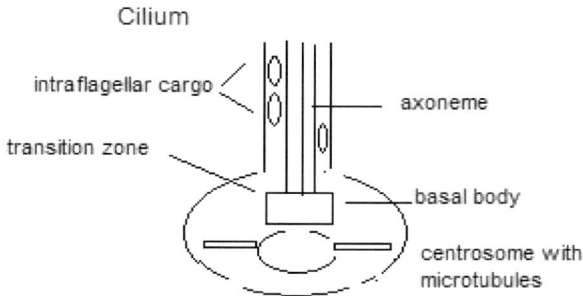

Fig. 6.3:   Cilium and transport.

associated with abnormalities in cerebral folding and structure in lissencephaly and corpus callosum dysgenesis.

Intraflagellar transport complexes A and B are multi-subunit protein complexes. Molecular motors including kinesin 1 and dynein 1 power the movements of these complexes.

BBSome

The BBSome (named because of defects in this complex in Bardet–Biedl syndrome) plays a key role in transporting proteins into and from cilia. The BBSome is located in the basal body of the cilia. And it also binds to transported proteins. This includes trafficking membrane proteins including transport of receptors and G-protein complexes.

Bardet–Biedl syndrome is characterized by manifestations that include retinitis pigmentosum, kidney failure, obesity, kidney failure, cognitive impairments, and skeletal malformations. At least 17 proteins are defined as BBS components, seven of these proteins cluster together and are stabilized by the interacting protein BBIP1.

In addition, at least three different proteins act to facilitate assembly of the BBSome. Cargo of the BBSome includes sonic hedgehog (SHH) signaling components and somatostatin receptors coupled protein receptor.

Scheidecker *et al.* (2014) reported that although 17 BBSome genes have been identified, 20% of patients with the Bardet–Biedl syndrome still lack molecular diagnosis. Through exome sequencing studies in a patient with this syndrome they identified compound heterozygous mutations in the *BBIP2* gene that encoded BBSome interacting protein that plays a role in BBSome assembly.

Nephrocystins

Proteins that cluster in the cilia transition zone close to the cell membrane include nephrocystin encoded by gene *NPHP1-11*. Mutations in the latter genes are also involved in nephronophthisis that may occur as a manifestation of the Bardet–Biedl syndrome. Cysts in the renal cortico-medullary junction characterize nephronophthisis. Mutations in 11 different genes can lead to this disorder. NPHP1 and nephrocystin mutations lead to juvenile onset disease. Defects in NPHP2 (inversin), may lead to infantile nephronophthisis with or without situs inversus. The protein inversin associates with beta tubulin, a ciliary axoneme component.

In humans KIF3a and KIF3b mutations have also been found to lead to kidney cysts and retinal dystrophy.

Polycystins encoded by the autosomal recessive and the autosomal dominant polycystic kidney disease loci are located in the basal body. Hildebrandt *et al.* (2011) reported that gene mutations in the *PKD1* and *PKD2* genes predispose to polycystic kidney disease but that cyst formation is likely due to rare somatic mutations (second hit mutations) in the adult polycystic kidney disease (*APKD*) genes. Polycystin-1 has been shown to play an important role in renal tubule morphogenesis.

Autosomal recessive polycystic kidney disease is also often associated with bile duct disease. The gene *PKHD1* encodes a protein referred to as fibrocystin or polyductin. Hildebrandt *et al.* (2011) reported that this protein is found in primary cilia in renal epithelial cells and there it colocalizes with *PKD2* encoded protein polycystin-2.

Joubert syndrome is a cilia malfunction syndrome due to mutations in any one of 21 different genes. Romani *et al.* (2013) reported that most proteins abnormal in Joubert syndrome cluster in the basal body or the transition zone of cilia. These genes include transmembrane protein encoded by the *TMEM67* gene, Tectonic genes, *TCTN1-3*. Other genes may also be involved in Joubert syndrome. In Joubert syndrome brain structural abnormalities, and retinal colobomas occur and patients may also manifest nephronophthisis. Defects in a number of different genes that encode proteins related to ciliary function can lead to this syndrome. These proteins include AHI1 (Abelson helper integration), nephronopthisis proteins (NPHP3, NPHP6, NPHP8), centrosomal protein (CEP290), and inositol polyphosphate phosphatase (INPP5E).

It is interesting to note that defects in several of the above genes may also lead to Meckel syndrome that is characterized by meningomyelocele, microphthalmia, lung hypoplasia and polycystic kidneys.

Hildebrandt *et al.* (2011) noted that studies of ciliary genes and developmental defects have revealed that different combinations of recessive mutation in genes that encode ciliary proteins can lead to a wide variety of different syndromes.

Motile cilia defects

Motile cilia defects lead to primary dyskinesia, associated with impaired respiratory function, impaired sperm motility and situs inversus, i.e., altered left to right body assembly and asymmetry. Defects in dynein encoding genes have been observed and also defects in assembly of dynein complexes, or defects in the components of the radial spokes within the axoneme.

A regulatory complex that determines the interactions between dynein protein and nexins have also been identified and found mutated in patients with ciliary dykinesia. Nexin attachments facilitate co-ordinated movement of the microtubule doublets in the axoneme. Defects in ciliary motility due to defects in dynein genes *DNAH5* and *DNAH11*, and DNAI1; mutations have been reported to lead to recurrent sinusitis, bronchiectasis and infertility.

Planar Cell Polarity

Planar cell polarity refers to cells positioned with apical-basolateral axis (see Figure 6.4). Planar cell polarity plays a role in cell migrations and in cell layering and intercalation (Simons and Mlodzik, 2008).

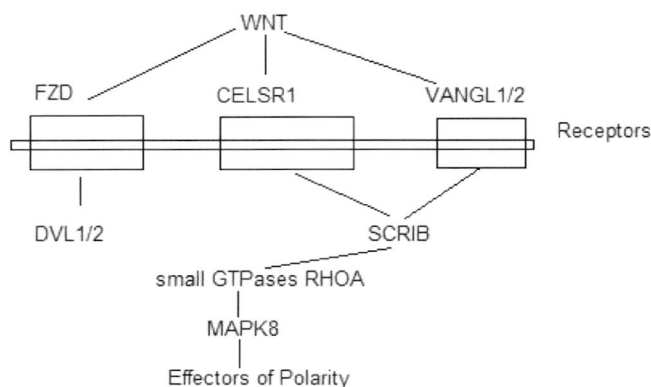

Fig. 6.4:   Planar polarity pathways.

Cell polarity and cilia development are connected and cilia development is connected with the centrosome. Planar cell polarity in vertebrates has been particularly well studied in the inner ear, in the organ of Corti, where sensory cells with stereo-ciliary bundles form rows.

The protein inversin that has similar domain architecture to the well-known cell polarity protein, Diego, is mutated in humans with ciliopathy nephronophthisis type 1 (Simons and Mlodzik, 2008).

CHAPTER 7

BRAIN DEVELOPMENT

Overview and Timing of Development

Tau and Peterson (2010) delineated six key periods in human embryonic brain development. During weeks 2 and 3 neurulation occurs, this process that involves formation of the neural plate and folding and fusion of ectoderm to produce the neural tube. In week 4, three vesicles develop, in the rostral neural tube. These vesicles include the forebrain, midbrain and hindbrain vesicles. The forebrain divides and separates to form the telencephalon that gives rise to the cerebral cortex and the diencephalon that gives rise to the thalamus and hypothalamus.

During weeks 5 and 6, the lining of the cerebral vesicles form the ventricular zone, the germinal matrix that gives rise to neuroblasts. In week 8, neuroblasts differentiate to specific neural cell types and microglia. Early steps in brain development include expansion of the neuro-epithelium and generation of radial glial cells in a region that subsequently becomes the sub-ventricular zone. Radial glial cells undergo specific asymmetric cell divisions to give rise to differentiated cells and progenitor cells. Radial glial cells establish connection that stretch from the sub-ventricular zone to the surface of the brain cortex.

During weeks 12–20 post-mitotic cells from the ventricular zone migrate along the radial glia. Neurons originating from the proliferative layer migrate tangentially toward the cerebral cortex and thalamus and give rise to GABAergic interneurons (gamma-amino butyric acid producing interneurons). By weeks 26–30 neuronal migrations are complete.

The development of neuronal circuits depends on extension of axons and dendrites from neuronal cell bodies and connection of the extensions with other partners. Tau and Peterson (2010) emphasized the importance of cytoskeletal proteins and scaffolding proteins in these processes. More mature circuits later replace early synaptic connections.

Neurulation, neural tube closure and neural tube defects (NTD)

Aspects of neurulation and neural tube closure and defects in humans were extensively reviewed by Copp *et al.* (2013). They reported that in humans neural tube closure begins at day 17 or 18 post-fertilization. Primary neurulation involves initial closure at the rostral (cranial) end of the neural plate fold, and closure occurs at several points along the neural plate fold. Secondary neurulation occurs at the lowest end of the neural plate fold.

Intensive research efforts on factors involved in neural tube closure in mouse and in humans have revealed that folate and one carbon metabolism play important roles. However Copp *et al.* (2013) stressed that in human the majority of cases of NTD are not due to folate deficiency. The exact mechanisms through which folate enhances neural tube closure are not known. However, folate and the one carbon metabolism are important for the synthesis of purines and pyrimidines and DNA, and are therefore important for cell proliferation and growth. In addition, folate and the one carbon pathway generate methyl groups that are important in histone and DNA methylation and epigenetic mechanisms of gene regulation.

There is evidence from mouse and human studies that the planar cell polarity pathway plays important roles in neural tube closure. This pathway includes non-canonical WNT signaling through receptors including FZD (Frizzled), CELSR (cadherin EGF LAG) and VANGL1 (planar cell polarity protein and downstream molecules important in this pathway include SCRIB (scaffold protein in planar cell polarity).

Neural tube closure defects

Copp *et al.* (2013) defined NTD as neurulation defects that included cranial defects, anencephaly, distal defects, spina bifida and craniorachischisis where there was extended closure failure. In closed NTD skin covers the defects, however, the spinal cord may be abnormally tethered to surrounding structures. They emphasized that these defects must be distinguished from herniation defects where defects in the skull or vertebral column lead brain or spinal cord to herniate through bone causing encephaloceles or meningomyelocele.

Genetic and environmental factors in the etiology of NTD

There is evidence that polygenic factors play important roles in NTD (see Table 7.1). In most cases these defects occur sporadically, however, following the birth of one child with NTD, the chances that a second child with NTD will be born to those same parents are higher, (2–5%) than the overall populations risk (0.1%).

Genes that encode proteins involved in folate and one carbon metabolism have been extensively studied. Amorim *et al.* (2007) carried out meta-analysis of the methylene tetrahydrofolate reductase variant MTHFR 677C>T rs1801133 and NTD. They reported that the associated TT genotype has only been proven in the Irish population.

There are other studies implicating folate and one carbon metabolism related gene variants in NTD, however, the sample sizes have been relatively small. Other genes implicated include MTHFD1, (methylene tetrahydrofolate dehydrogenase) and the mitochondrial located counterpart MTHFD1L.

Narisawa *et al.* (2012) reported that mutations in the glycine cleavage system (see Table 7.1) predispose to NTD in mice and humans. The glycine cleavage system decarboxylates glycine in a complex reaction and generates 5–10 methylene tetrahydrofolate from tetrahydrofolate. Mice with impaired glycine cleavage developed NTD. The defects were not preventable with folate but were preventable with methionine.

There is evidence that low levels of choline and phosphatidylcholine are associated with an increased frequency of NTD. Choline acts as a methyl donor. Specific mutations in the enzymes BHMT or BHMT2, (betaine-homocysteine-S-methyltransferase) enzymes involved in choline metabolism, have been associated with NTD. Specific defects in cubulin (CUBN) and transcobalamin (TCN2) that are involved in Vitamin B12 metabolism have been reported in some cases with NTD (Imbard *et al.*, 2013).

Transport of folate in the brain

Insight into folate transport in the brain was gained in part through studies on patients with cerebral folate deficiency. Manifestations of this syndrome include ataxia, dystonia, seizures, developmental delay and visual

Table 7.1:   Gene products mutated in Neural Tube defects.

(A)	**One Carbon Metabolism**

Protein Symbol	Protein Function
SHMT	Serine hydroxymethyltransferase
MTHFR	Methylene tetrahydrofolate reductase
BHMT	Betaine — homocysteine S-methyltransferase
BHMT2	Betaine — homocysteine S-methyltransferas2
CUBN (cubulin)	Cobalamin receptor (intrinsic factor)
TCN2	Binds and transports Cobalamin (Vitamin B12)
SLC46A1 (PCFT)	proton-coupled folate transporter
SLC19A1 (RFC1)	intra-cellular folate transporter
FOLR1/FOLR2	membrane bound folic acid binding proteins

(B)	**Glycine Cleavage System**

Protein/Gene Symbol	Protein Function
GLDC	Glycine dehydrogenase decarboxylating
P protein	Pyridoxal phosphate dependent glycine decarboxylase
H protein	Lipoic acid containing protein
T protein	Tetrahydrofolate requiring enzyme
L protein	Lipoamide dehydrogenase

(C)	**Planar Cell Polarity Non-Canonical WNT Pathway**

Protein/Gene Symbol	Receptors
FZD	WNT receptors
VANGL1/2	Planar cell polarity protein receptors
CELSR1	Cadherin EGF LAG G protein receptor

(D)	**Downstream Effectors**

DVL1/2	Disheveled segment polarity protein
SCRIB	Scribbled protein polarity protein
PTK7	Protein kinase 7
RHOA	Ras homolog actin cytoskeleton regulator
MAPK8 (JNK)	Map kinase

problems. The key laboratory finding is the low level of 5-methyltetrahydrofolate in the cerebrospinal fluid (CSF).

Entry of folate into the brain requires that it crosses the blood brain barrier (BBB). This barrier is formed by capillary endothelial cells, by proteins expressed in membranes of those cells and by tight junction proteins that occur between those cells (Wang *et al.*, 2013). They reported that the BBB forms in the embryo by the end of the first trimester and that it is fully functional shortly after that time.

Folate transport proteins include ATP binding cassette protein a P-glycoprotein transporter proton coupled folate transporter PCFT (also known as SLC46A1), folate transporters RFC1 (SLC19A1). Folate receptors FOLR1 and FOLR2 play important roles and participate in folate mediated receptor endocytosis. FOLR1 is located on epithelial cells and FOLR2 is primarily expressed on mesenchymal cells. Following binding of folate to the receptors invagination of the cell membrane occurs leading to endosome formation at the membrane cytoplasmic interface. Within the endosome compartment folate is released. Zhao and Goldman (2013) reported that the transporter SLC46A1 likely to play role in the release of folate from the endosome. They emphasize that SLC46A1 and FOLR1 are required for transport across the choroid plexus epithelial cells.

The PCFT folate transporter (SLC46A1) is expressed in liver, kidney, lung, heart, brain and spinal cord. Wang *et al.* (2013) determined that in brain capillaries, PCFT colocalizes with the P glycoprotein, PCFT mediates transport across the choroid plexus to the CSF. They demonstrated that in PCFT deficient mice the tight junctions that exist between epithelial cells were also disturbed. Mutations in folate receptors have been found in patients with primary folate deficiency syndrome.

Autosomal recessive SLC46A1 defects in humans lead to a condition that manifests within a few months of birth and is characterized by anemia, immune deficiency and absence of CSF folate. Subsequently patients develop dystonia and seizures, the manifestations of cerebral folate deficiency syndrome. Treatment includes high doses of oral formyl tetrahydrofolate.

Wang *et al.* (2013) reported that impaired brain folate transport and impaired tight junction protein function occur in a number of neurologic and metabolic conditions including disorders characterized by impaired mitochondrial function. These conditions lead to secondary folate deficiency syndrome.

The form of folate that is primarily delivered from blood to tissues is 5-hydroxymethyltetrahydrofolate (5HMTHF). SLC19A1 (also known as RFC1) is a key transporter. SLC1 is expressed in the apical brush border membrane of the intestine, in the choroid plexus, retinal pigment epithelium, in the basolateral membrane of the renal tubule and in the placental syncytiotrophoblasts (Zhao and Goldman, 2013).

Folate deficiency and abnormal folate metabolism may also arise secondary to defects in serine metabolism and in glycine metabolic defects.

Herniation defects and genetic risk

Copp *et al.* reported that the herniation defect encephalocele occurs in Meckel syndrome due to impaired cilia function (see the section on organelles and cilia). Meckel syndrome arises due to defect in any one of a number of genes that play roles in cilia, TMEM216, TMEM67 (involved in the formation of primary cilia, centrosomal protein CEP290 and centrosome/cilia gene RPGR1PIL).

Environmental factors and NTD

Fungal contaminations of food and toxin production (fumonisin) were reported to increase the incidence of NTD (Copp *et al.*, 2013). There are also reports of increased frequency of neural tube defects in fetuses exposed to anti-convulsant medications (valproic acid, carbamazepine) and in association with maternal obesity.

Analysis of brain corticogenesis from stem cells

A number of studies on cultures of embryonic stem cells (ESC) have revealed intrinsic pathways that lead to the generation of the cerebral cortex. Gaspard *et al.* (2008) cultured ESC at low density in chemically defined medium in the absence of growth factors and in the presence of insulin. They determined that the ESC underwent neurogenesis in a stepwise manner. First, neural progenitors appeared and were defined as nestin positive cells. These then generated neuron precursors that were positive for MAP2 (microtubule associated protein) and tubulin. Culture of neural progenitors in the presence of inhibitors of SHH (sonic hedgehog gene product) was carried out to promote patterns of forebrain development. Under these conditions the neural progenitors developed characteristics of pyramidal neurons and were positive for the glutamate transporters, SLC17A7 and SLC17A6.

Pyramidal cells are classified according to the markers they express and ultimately through their positions in the layered cerebral cortex. In the ESC cultures, Gaspard *et al.* (2008) identified reelin positive cells which are the Cajal Retzius cells that form the earliest cortical neurons. Subsequently they identified TBR1 positive cells and sequentially neurons were identified that expressed the transcription factors OTX1, CIP2, SATB2 and CUX1 (see Table 7.2). Specific types of pyramidal cells with defined marker expression patterns predominated in particular layers that developed. Furthermore cells within specific layers projected their axons differently.

Table 7.2:   Proteins in Brain Development.

(A) Neuron Differentiation

Protein/Gene Symbol	Protein Function
TBR2 (Eomes)	Transcription factor for mesoderm and brain
OCT4 (POU5F1)	Transcription factor pluripotency marker
NANOG	Homeobox, brain, pluripotency markers
TFAP2 (AP2gamma)	Transcription factor developmental gene activator
INSM1	Neuroectoderm differentiation (in insulinomas)
CUX1	Cut-like homeobox, DNA binding, regulation
SATB2	Transcription regulation and chromatin remodeling
FEZ2	Axon bundling and elongation
CTIP2 (BCL11B)	Zinc finger protein, transcription repressor
SOX5	In protein complex, regulates transcription
NES (Nestin)	Intermediate filament in nerve cells
MAP2	microtubule associated, in dendrites
TUBB (Tubulin)	Structural component of microtubules
SHH (sonic hedgehog)	Neural tube patterning
SLC17A6	Vesicular glutamate transporter
SLC17A7	Vesicular glutamate transporter

(Continued)

Table 7.2:   (*Continued*)

RELN (Reelin)	Extra-cellular matrix, cell-cell contact
OTX1	Homeobox, brain and sensory organ
CDH2 (NCAD)	Presynaptic post-synaptic adhesion
SOX1	Transcription factor, neural induction
PAX6	Transcription factor, neural induction
FOXG1	Forkhead transcription repressor
SIX3	Homeobox transcription factor
PAX2	Homeobox, transcription factor
GBX2	Gastrulation homeobox
EGR2 (KROX20)	Early growth response transcription factor
AUTS2 (autism associated)	Neurodevelopment
TSHZ2	Zn finger homeobox
CDK5RAP2	Cyclin dependent kinase (cell division)

Neuronal Cell Markers

Parvalbumin (basket cells and interneurons)

Cholestokinin (basket cells)

Somatostatin (interneurons)

Neuropeptide Y (serotonin receptor interneurons)

Migration of Interneurons

Neurotrophins

Semaphorins

SLIT and receptor ROBO

Neuregulin

Glia Oligodendrocyte Differentiation

PAX6 and SOX1

OLIG2 (oligodendrocyte lineage transcription factor)

Myelin Sheath

Phospholipoprotein (PLP),

Myelin basic protein (MBP)

2'3 cyclic nucleotide 3 phosphodiesterase (CNP)

Myelin associated glycoproteins (MAG).

(*Continued*)

Table 7.2: (*Continued*)

NKX2.2 (homeobox transcription factor)

SOX10 transcription factor

PDGFRA (platelet derived growth factor receptor).

Neuronal Positioning, Axon Guidance

Reelin and lipoprotein receptors

DAB1(reelin signal transducer)

Netrins

DCC DSCAM Netrin receptors

Transsynaptic Adhesion Molecules

Neurexins

Neuroligins

Syncams

Post-Synaptic Density

DLG1-DLG4

Shank Proteins

AKAP protein kinase

HOMER, regulator of neurotransmitter function

Lancaster *et al.* (2013) described a three dimensional system for culturing embryonic stem cells; this system gave rise to brain organoids. Their system is built on evidence for the self-organizing capacity of pluripotent stem cells to form tumors. They maintained neuroectodermal cells in Matrigel scaffold and then in a spinning bioreactor. Under these conditions cerebral organoids were generated, they were initially characterized as N cadherin positive neuroepithelial structures that surrounded a fluid filled cavity. During the course of differentiation within the organoids the pluripotency markers OCT4 and NANOG decreased and the expression of neural induction markers SOX1 and PAX6 transcription factors increased. Thereafter, both forebrain markers FOXG1 and SIX3 (transcription factors) were expressed as well as hindbrain markers EGR2 (earth growth response protein transcription factor) and ISL1 (Lim homeobox transcription factor). Subsequently hindbrain markers decreased and forebrain expression markers predominated.

Lancaster *et al.* (2013) documented that cerebral cortical layering was developed in organoids using markers TBR1 (T-box brain transcription factor) and CTIP2 and SATB2 transcription regulators. They noted evidence for brain regionalization as the organoids developed further. They documented expression of AUTS2 a prefrontal cortex marker and TSHZ2 (a zinc finger homeobox transcription factor), an occipital lobe marker.

Lancaster *et al.* (2013) proposed that organoid culture provides a novel approach to understanding neurodevelopmental disorders. In their studies on organoids derived from stem cells established from a child with microcephaly due to mutation in the gene CDK5RAP2 (CDK5 regulatory sub-unit associated protein). They determined that the mutation led to premature neural differentiation of early progenitors and failure of continued progenitor cell generation.

Neuronal Maturation, Outgrowth of Dendrites and Axons

Maturation of neurons requires the activity of products of proneural genes, e.g., neurogenins NEUROG1, NEUROG2 and basic loop helix transcription factors, e.g., ASCL1 transcription factor. Ali *et al.* (2014) reported that post-translation modifications, e.g., phosphorylation of proneural proteins were necessary for maturation of CNS progenitor stem cells.

Cell division, microtubules neuronal migration and brain size

Remodeling of microtubules precedes cell division. At the start of prophase microtubules surround the nucleus. Later in prophase when centrioles have replicated and moved to opposite poles of the cell microtubules extend from the centrosomes at the poles and form the spindle at the center of the cell. Astra microtubules radiate out from the polar centrioles toward the cell membrane. Microtubules also connect the spindle with the kinetochores at the centromeres of the chromatid pairs. At the end of prophase, the nuclear membrane disintegrates. During anaphase the chromatid pairs within each chromosome separate and move to opposite poles. Microtubules constitute the scaffold and track along which the chromatids move during mitosis.

Microtubules are composed of heterodimers of alpha and beta tubulin sub-units. Each heterodimer binds a GTP (guanosine 5'triphosphosphate). Microtubule associated proteins include MAP1A, MAP1B that occur in dendrites and axons. MAP2A and MAP2B also occur in dendrites, and MAP2C occurs in embryonic dendrites, and MAP4 occur in neuron cells bodies. TAU is a microtubule binding protein that occurs in axons and dendrites. The binding of microtubule associated proteins to microtubules is mediated by MAP kinase.

In the nervous system, microtubules play important roles in transporting nutrients and vesicles. Microtubule associated proteins are important in neuronal migration and in of synaptic vesicles trafficking. Specific motor proteins that facilitate transport include dynein, dynactin and kinesin motor proteins. KIF1A is a kinesin motor protein that transports synaptic vesicles. Kinesin motor proteins also play roles in trafficking proteins at the synapses. Liu *et al.* (2012) reported that there are at least 45 genes in the human genome that regulate microtubule trafficking.

Cortical Malformation Syndromes and Defects of Cell Cycle and Microtubule Dynamics

Studies on human cortical malformation syndromes including microcephaly and lissencephaly have yielded insights into the importance of cell cycle dynamics, centrosome and microtubule related mechanisms in cortical development.

Primary microcephaly

This is defined as head circumference of at least two standard deviations below normal at birth and the degree of microcephaly is often greater later in life. In primary microcephaly there are usually no associated brain malformations. Primary microcephaly may be inherited as an autosomal recessive condition (Dyment *et al.*, 2013).

Gilmore and Walsh (2013) reported that non-syndromic primary microcephaly in humans is the most frequent due to mutations in genes that have functions related to cell division mechanics including centrosome biogenesis and integrity, mitotic spindle organization and microtubule organization. Genes that are mutated in specific forms of microcephaly include ASPM1 (abnormal spindle microcephaly associated), CASC5 (microtubule attachment and chromosome segregation), WDR62 (WD repeat protein) centrosomal proteins CEP152, CEP63, CENPJ. In addition, mutations in CDK5RAP2 binds to kinase regulator, and in STIL, a mitotic spindle checkpoint protein is associated with microcephaly. Other genes found mutated in microcephaly include genes involved in DNA damage detection and repair, RAD50, ATR (cell cycle checkpoint), NBN (nibrin double stranded DNA break repair).

Chromatin regulation, neurogenesis and microcephaly

Yang *et al.* and Walsh (2012) emphasized that both symmetric and asymmetric division of cerebral cortical progenitor cells are required for brain development and maintenance. They noted that the molecular processes that regulate these divisions are not well defined. However, there is evidence that chromatin modifications and remodeling and the Polycomb and Trithorax group complexes play important roles. REST silencing transcription factor also impacts epigenetic regulation of neurogenesis. REST acts in part by repressing expression of specific genes through recruitment of histone deacetylases.

Yang *et al.* (2012) reported discovery of a new regulator of vertebrate neurogenesis, ZNF335. This regulator was discovered through genomic linkage and DNA sequencing studies in a family where several children had very severe microcephaly. In further studies they revealed that ZNF335 normally interacts with histone 3-lysine 4 (H3K4) methyltransferase complex. Furthermore they demonstrated in mice that knockdown of ZNF335 disrupts neuronal progenitor cell proliferation and differentiation.

In the specific microcephaly cases in the family, Yang *et al.* (2012) studied there was a homozygous mutation in ZNF335 that disrupted a splice site. This resulted in reduced production of gene product. They determined that normal ZNF335 interacts with the Compass complex that is a component of the trithorax TRXG complex that impacts H3K4 methyltransferase activity. They determined further that ZNF335 bound to the promoter region of REST. Reduction of ZNF335 led to the decreased REST expression. Loss of ZNF335 reduced levels of REST, COREST and also altered levels of other key genes involved in brain development, including neurogenin, oligophrenin and NF1B nuclear factor. They noted that the histological studies revealed that ZNF335 knockdown revealed not only evidence of reduced neuronal cell proliferation and differentiation but also evidence of neuronal degeneration. Studies on the molecular basis of severe microcephaly in a family and discovery of ZNF335 therefore provided further insight into the function of the trithorax complex in neurogenesis.

Neuronal migration defects and aberrant brain cortex structure

The protein doublecortin (DCX) and doublecortin like kinase 1 (DCLK1) both impact neuronal migration and lead to aberrant cerebral cortex structure. DCX mutations lead to sub-cortical bands heterotopias. Liu *et al.* (2012) reported that in these disorders the kinesin motor protein KIF1A is mislocalized and its function is impaired. KIF1A and DCX form a complex on the microtubule.

A number of genes that encode products involved in neuronal migration were identified through genetic studies in patients with lissencephaly. In lissencephaly the brain surface is smooth, cortical gyri are abnormal and the cortex is composed of four layers and not the normal six layer structure (Reiner and Sapir, 2013).

In some cases lissencephaly is due to deletion in the LIS1 gene on chromosome 17p13.3. Large deletions in this region may involve LIS1 and flanking genes and lead to Miller Dieker syndrome characterized by severe brain malformations and facial dysmorphology. In some cases somatic mutation and deletions occur only in a proportion of neurons and this may result in band heterotopia or cortical heterotopia (Pagnamenta *et al.*, 2012).

It is interesting to note that LIS1 deletions lead to lissencephaly while duplication of the LIS1 gene leads to microcephaly without overt lissencephaly (Lockrow, 2012). There is evidence from gene knockdown experiments in mice that LIS1 plays a role in proliferation of neurons and astrocytes and in cell migration. LIS also interacts with the NDE1 protein that is centrosome associated and plays a role in neuronal migration.

SHH pathway and brain deformities

The SHH pathway plays a key role in proliferation of neural stem cells in brain. SHH mutations in humans lead to midline deformities involving the midbrain and face and lead to holoprosencephaly (Roessler *et al.*, 1997). In this condition, there are failures or the embryonic forebrain (prosencephalon) to divide into double lobes of the cerebral hemisphere. In severe forms a single cortical lobe may occur. Komada (2012) reported that SHH signaling regulates cell cycle kinetics of radial glial cells and neural progenitor cells to maintain proliferation

Development of axons, dendrites and synapses

There is evidence that the WNT signaling pathway is involved in axon path finding, in dendritic development, in the formation of synapses and in the maintenance of synaptic plasticity (Budnik and Salinas, 2011). In humans there are 19 different genes that encoded WNT family proteins. In addition, proteins in at least three different families can form WNT receptors.

Components of the post-synaptic density

Dendritic spines are laden with post-synaptic receptors. A dense protein rich structure underlies the tip of the dendritic spine and is referred to as the post-synaptic density (PSD). Scaffolding proteins constitute important structural components of the PSD (Ting *et al.*, 2012).

Scaffolding proteins include membrane associated guanylate kinase proteins (MAGUK), Synaptic associated proteins, (SAP) DLG 1-DLG4 (Discs large), SHANK proteins (ankyrin and repeat domains for protein binding) and AKAP proteins. The AKAP proteins are anchor protein kinases that regulate the activity of protein kinase A. Also included in the PSD are Homer proteins that occur in long and short isoforms and regulate neurotransmitter receptor function. Long isoforms link together with other proteins of the PSD including SHANK proteins and control the PSD structure. Metabotropic glutamate receptors and ryanodine proteins also bind to the PSD protein Homer.

Three genes encode SHANK proteins, the multiple ankyrin repeat proteins. SHANK proteins interact with the SAP proteins and Homer proteins and play roles in PSD stabilization.

Mutations in genes involved in the production of PSD proteins occur in a number of psychiatric and behavioral disorders (Ting *et al.*, 2012).

DLG mutations

The PSD protein encoded by DLG1 is deleted in the chromosome 3q29 microdeletion syndrome that is characterized by mild to moderate mental retardation, ataxia and autism; patients also manifest facial dysmorphology. DLG1 copy number variations were described in patients with schizophrenia and DLG4 mutations were reported in patients with schizophrenia and patients with autism (Cheng *et al.*, 2010). DLG4 maps to the X chromosomes and mutations in this gene occur in some patients with X-linked mental retardation.

Mutations or deletions in the DLG associated protein DLGAP3 may play a role in generation of obsessive-compulsive behaviors. Cheng *et al.* (2010) reported that mice with deletions in Dlgap3 (Sapap3) had increased anxiety and compulsive self-grooming. Furthermore, these mice had defects in glutamatergic transmission at the cortico-striatal junction. These defects could be eliminated by introduction of Sapap3 in the striatum.

Neural cell types

Different methods have been applied to group neurons into functionally related groups. Descriptions of cells were used, (Nelson *et al.*, 2006). In earlier studies morphological descriptions of cell were used. Nelson *et al.* reviewed more recent classification methodologies for more than 60 different neuron cell types that are currently defined in the retina and for the thousands of cell types that exist in brain. Methodologies applied to analysis of cortical and hippocampal neurons include electrophysiology, immuno-cytochemistry and gene expression analysis. Nelson *et al.* (2006) emphasized that even within a specific cell type there are frequently a range of different cells. One example presented was the neocortical fast-spiking parvalbumin positive basket cells, multiple sub-types of these cells occur. Parvalbumin positive basket cells represent a subclass of Gaba-ergic neurons.

Toledo-Rodriguez *et al.* (2005) classified neurons on the basis of expression of neuropeptides, including neuropeptide Y, vaso-active intestinal peptide, somatostatin and cholecytokinin and calcium binding proteins including calbindin, parvalbumin and calretinin. They determined that a strong correlation existed in the expression of specific combinations of neuropeptides and calcium binding proteins.

Increasingly microarray analyses of gene expression patterns are used to define different cell types. Greig *et al.* (2013) reported that early in corticogenesis undifferentiated neurons overlap in expression. As development proceeds neurons differentiate to more distinct sub-types. They demonstrated that this differentiation was in part driven by expression of key regulators that were either activated or repressed in different cell types. Key regulators included SATB2 (special AT sequence binding protein, TBR1 (T box brain), SOX5 (SOX family transcription factor), CTIP2 (BCL11B) (a BCL related transcriptional repressor), FEZF2 (forebrain embryonic zinc finger-like protein.

In Gaba-ergic neurons, the amino-acid glutamate is converted to gamma amino-butyric acid through the activity of glutamate dehydrogenase (GAD). In the brain GAD1 (GAD67) is particularly important in this reaction in cortical regions, including dorso-lateral prefrontal cortex, sensory, motor and limbic regions.

Lewis *et al.* (2011) reported that levels of GAD1 are reduced in the dorso-lateral frontal cortex of patients with schizophrenia and that the GAD1 deficiency was particularly marked in the parvalbumin positive interneurons and parvalbumin positive basket cells. In addition, reduced expression of cholecystokinin from cholecystokinin positive basket cells was noted in schizophrenia.

Cortical interneurons

Wonders *et al.* (2007) reviewed the anatomical origin of cortical interneurons and also the occurrence of sub-groups of these interneurons. Cortical projection neurons derive from the telencephalon and undergo radial migration to the cortical mantle. The interneurons have somewhat different sites of origin and they migrate tangentially in the developing brain.

GABAergic interneurons are inhibitory neurons that produce and release GABA (gamma aminobutyric acid). Different sub-types of these neurons exist. One sub-type of GABA producing interneurons has short projecting axons and smooth or sparse spine dendrites. GABAergic interneurons are classified on the basis of the cell surface markers they express including parvalbumin, somatostatin and serotonin receptors.

Marin (2012) reported that more than 20 different classes of interneurons have been identified in the cerebral cortex and these differ from interneurons in the amygdala and striatum. The different sub-types of interneurons express different transcription factors and those in turn determine expression of proteins that define sub-type characteristics including ion channel composition, migration and position.

In recent years substantial progress has been made in mapping the transcription factor networks that determine cortical neuron sub-types. Kelsom and Lu (2013) reported that 40% of the GABAergic cortical interneurons express parvalbumin and have a fast spiking firing pattern. With respect to morphological characteristics the interneurons that express parvalbumin include basket cells and chandelier cells. Kelsom and Lu noted that the fast-spiking basket neurons play important roles in regulating the balance between excitatory and inhibitory inputs.

Somatostatin expressing interneurons are sometimes referred to as Martinotti cells. However, a number of different cell types exist in the somatostatin interneuron category.

Interneurons in the serotonin receptor category differ in morphology and have multiple dendrites. These interneurons may interact with each other and with other interneuron cell types. They frequently express neuropeptide Y.

Kelsom and Lu (2013) reported that origins and migration routes of cortical interneurons differ in rodents and primates. In the developing human brain there may be multiple sources of cortical interneuron progenitors. Specific factors that impact the migration of interneurons include growth factors, neurotrophins and glial derived neurotrophic factors. The direction of migration is also impacted by chemo-repellents such as semaphorins. SLIT1 is a ligand for the ROBO 1 receptor that plays a role in axon guidance. Ephrin receptor kinases also act as chemo-repellents. Chemo-attractant molecules also determine the migration patterns of interneurons; neuregulin 1 encoded by the *NRG1* gene is an example.

Neurotransmitter systems

Neurotransmitters include aminoacids such as glycine and glutamine, or molecules synthesized in presynaptic terminals. The latter include acetylcholine, gamma amino-butyric acid, nor-epinephrine, dopamine and serotonin. Neurotransmitters depend on receptors for their activity. These receptors may be ion channel receptors or they may be coupled to second messenger systems.

Glial cell types

These include non-neuronal cells oligodendrocytes, microglia and astrocytes. Fields (2010, 2013) emphasized the important roles that glia play in brain function. Oligodendrocytes form myelin and provide metabolic support for axons. Astrocytes secrete extra-cellular matrix (ECM) proteins. They also form sheaths around synapses thereby impacting synaptic transmission.

Microglia are highly mobile cells. Fields reported that these cell monitor electrical activity and prune synaptic connections.

Studies of oligodendrocytes generated from progenitors have revealed that epidermal growth factor (EGF) signaling plays an important role in myelin generation. Aguirre *et al.* (2007) revealed that EGFR signaling impacted myelination of the corpus callosum.

Myelin sheath

The myelin sheath is composed of lipids, of myelin proteins and of extensions from oligodendrocytes. Myelin proteins include phospholipoprotein (PLP), myelin basic protein (MBP), 2'3 cyclic nucleotide 3 phosphodiesterase (CNP) and myelin associated glycoproteins (MAG).

Fulton *et al.* (2010), reported that PLP, and myelin basic protein (MLP), contribute to the multi-layered structure of myelin while MAG, are limited to the peri-axonal region and facilitate connections between the axonal membranes and myelin. They emphasized that the MBP and the PLP genes both give rise to alternate transcripts and that some of the isoforms have function unrelated to myelination. Protein isoforms generated from the PLP gene include PLP and DM20. DM20 is expressed primarily during embryonic development in non-myelinating cells. The PLP isoform manifests post-natal expression. PLP gene products likely play roles in processes such as ion exchange, cell migration and cell death. Fulton *et al.* (2010) emphasized that PLP serves as a membrane protein that participates in the transduction of signals from integrins in the extra-cellular matrix to the interior of the cell.

Two protein isoforms are derived from the CNP gene. CNP1 and CNP2 are both expressed in myelinating cells. CNP2 occurs primarily in mitochondria and is expressed in neural and non-neural tissues.

The large 100 kb MBP gene contains three different transcription start sites. Start sites 2 and 3 generate transcripts for MBPs that are major proteins of the myelin sheath and play roles in the compaction of myelin. Fulton *et al.* (2010) described the second class of MBP transcripts generated from transcription site 1 as golli proteins. These proteins expressed in many tissues are most likely involved in protein–protein interactions. Golli protein expression is high during embryonic life. There is evidence that golli proteins interact with voltage gated ion channels and that this interaction plays important roles in oligodendrocyte migration and in the extension of oligodendrocyte processes. Golli proteins also regulate calcium uptake through the store operated calcium channels (SOCC). MBPs interact with the cytoplasm in oligodendrocytes and bind to actin and tubulin. Fulton *et al.* (2010) noted that there is some evidence that MBP interacts with microtubules.

Stem cell analysis of oligodendrocyte differentiation and myelination

Wang *et al.* (2012) developed protocols to generate oligodendrocyte precursor cells from human induced pluripotent stem cells (hiPSC) and from human ESC. They assessed the competency of these cells to induce myelination in shiverer mice; these mice have myelin deficiency due to an autosomal recessive mutation in the Mbp gene.

Wang *et al.* (2012) devised a six-stage differentiation protocol for the cultured cells. The protocol extended between 110 and 150 days. Different factors were added at each stage to achieve transition of stem cells to oligodendrocytes. In stage 2, cells expressed transcription factors PAX6 and SOX1. In stage 3, oligodendrocyte precursor cells expressed the markers OLIG2 (oligodendrocyte lineage transcription factor) and NKX2.2 (homeobox transcription factor). With further differentiation cells expressed OLIG2, NKX2.2, SOX10, PDGFRA (platelet derived growth factor receptor). Stage 6 cells also expressed MBP and glial fibrillary acidic protein

(GFAP) and at that stage both astrocytes and oligodendrocytes were present in the cultures. Wang *et al.* (2012) then used fluorescence activated cell sorting to isolate oligodendrocyte precursor cells.

To investigate the myelination capacity of the stem cell derived differentiated oligodendrocytes, Wang *et al.* (2012) implanted these cells in the corpus callosum of the shiverer mice. They determined that these cells were to robustly myelinate fibers within the brain to promote survival of the animals. Proof of myelination of axon fibers was achieved through electron microscopy. Analysis of the marker CASPR (contactin associated protein) at the nodes of Ranvier in nerve fibers revealed normal structure. Nodes of Ranvier are periodic gaps in the myelin sheath that apparently facilitate transmission of the action potential along nerves. There was no evidence that the transplanted oligodendrocytes were tumorigenic.

Basement membranes and extracellular matrix (ECM)
in the nervous system

Basement membrane of endothelial cells, pericytes and astrocytes secrete ECM molecules. Obermeier *et al.* (2013) emphasized the importance of interactions of basement membrane molecules collagen type IV, laminin and fibronectin with glycosaminoglycans.

Barros *et al.* (2011) reviewed the key roles that ECM glycoproteins have in regulating neural stem cell functions, neuronal migration, formation and interactions of axonal processes and synapses. Radial glial cells represent specific forms of neural stem cells. The ECM that surrounds these cells modulates their maintenance and migration.

Key ECM proteins in the nervous system include laminins, reelin, chondroitin sulfate glycoproteins, perlecan, integrin, dystroglycan, tenascin, thrombospondin type 1 repeat proteins.

Laminins are found in contact with end feet of radial glial cells; laminins also surround blood vessels. Defects in laminins lead to defects in establishment of cortical layers. Patients with defects in dystroglycan glycosyltransferases manifest migration defects and neuronal ectopias. Radial glial cell migration defects and neuronal ectopias also occur in cases with mutations in alpha or beta integrins.

Barros *et al.* (2011) noted that reelin is one of the best studied ECM glycoproteins in the nervous system and it plays key roles in neuronal migration and cortical lamination. Reelin plays critical roles during development in neuronal positioning and migration and in cell–cell interactions. There is also evidence that reelin has different functions in different brain regions. Reelin mutations in humans lead to lissensephaly, characterized by absence of convolutions and folding of cortical gyri. Reelin binds to the receptor APOER2 low density lipoprotein related 8. Reelin also binds protein binds to the lipoprotein receptor VLDRL (very low density lipoprotein receptor) and this binding activates phosphorylation of the associated intra-cellular DAB1 protein by SRC kinases. DAB1 plays a critical role in axon guidance.

Barros *et al.* (2011) reported that extra cellular matrix proteins that include thrombospondin type 1 repeats play important roles in regulation of axon growth. Axonal outgrowth and guidance are also impacted by the secreted protein netrins and slits. Netrins are secreted by axonal target cells and they act primarily as chemoattractant molecules. Netrin receptors include DCC encoded on chromosome 18 and DSCAM a protein encoded on chromosome 21.

In mice defects in the corpus callosum and hippocampal commissure occur in mice with specific mutations in netrin or in the netrin receptors. Slit proteins are secreted glycoproteins expressed in the central nervous system, they bind to ROBO (Roundabout) receptors and play roles in the establishment of axonal tracts.

Barros *et al.* (2011) emphasized that networks of secreted proteins surround synapses. This network is composed of hyaluronic acid and chondroitin containing proteoglycans. Other components of this network include neurocan, versican and tenascin. Thrombospondins are astrocyte secreted glycoproteins that are important in synatogenesis and synapse maturation.

Synaptic plasticity, pruning and neuronal cell death

Tau and Peterson (2009) reviewed aspects of neural circuitry development and the role of pruning. They noted that pruning processes include elements of axons, dendrites and synapses. These processes begin in late gestation and become more active in the post-natal period.

They defined synaptic plasticity as the process through which activity dependent signals are modified at synapses. They emphasized the critical roles of synaptic remodeling and pruning in the thalamo-cortical projections and sensory systems. Of particular interest in their studies are the ways in which experience and environment impact neural circuits and functional connectivity.

Hyman and Yuan (2012) reported that caspases, that act as cysteine proteases play important roles in axon pruning and synapse elimination during development.

Kanamori *et al.* (2013) reported that calcium signaling plays a key role in dendritic pruning and neural circuit sculpting. Through their studies in Drosophila that determined that increased intrinsic excitability and locally increased calcium flux occurred in dendrites approximately three hours before branch elimination. They proposed that the calcium activated enzyme calpain acts downstream of calcium signaling.

Iasevoli *et al.* (2013) reviewed molecular mechanisms that influence synaptic plasticity. They noted that the PSD scaffolding proteins play important roles in determining plasticity through regulation of receptor localization. PSD scaffolding proteins connect receptors and their intra-cellular second messenger effectors, intra-cellular ion channels, small GTPases and actin cytoskeleton. Specific PSD proteins impact the degradation of receptors.

Trans-synaptic adhesion molecules connect the pre and post-synaptic components Missler *et al.* (2012) reviewed these molecules that include neurexins, neuroligins and immunoglobulin (Ig) domain molecules, e.g., syncams and proteins rich in leucine repeats. The cellular domains of these proteins are often rich in repeat units e.g., immunoglobulin-like domains, or cadherin domains. These repeat units increase possibilities for interactions with other proteins and also provide mechanical stability. There is evidence that neurodevelopmental disorders may result from disorganization at the synapse.

The Blood Brain Barrier

Abbott *et al.* (2010) reviewed the structure and function of the BBB. They stated that evolutionary pressure for development of this barrier derives from the importance of establishing appropriate conditions for chemical and electrical functioning of synapses and axons that require precise regulation of their microenvironments. The major site of the BBB lies in the microvessel endothelium. Another important structure in BBB function includes the choroid plexus epithelial cells across which the CSF is secreted. The arachnoid membrane encloses the brain. This membrane is relatively avascular.

In their review of development of the BBB, Obermeier *et al.* (2013) emphasized the unique characteristics of the cerebral vessel endothelial cells. They differ from endothelial cells in other tissues in having specific connecting intercellular tight junctions and few fenestrations. These features limit transcellular and paracellular molecular movements. Specific solute transporters regulate molecular passage across the brain endothelial cells. Obermeier *et al.* (2013) documented the major signaling pathways involved in the development of the BBB. The first step involved invasion of endothelial cells into the neuroectoderm in response to vascular endothelial growth factor secreted by neural progenitor cells and guided by FLK1 (KDR receptor tyrosine kinase) and GPR124 (a G protein guided receptor expression). FZD receptor is expressed on endothelial cells and this interacts with WNT1 expressed by neural progenitor cells.

During the next stage in BBB development pericytes and astrocytes migrate toward endothelial cells and adhere to them. This recruitment is enhanced through expression of platelet derived growth factor PDGF beta from epithelial cells and PDGF beta receptor on pericytes. In addition, transforming growth factor TGF beta and TGFbeta receptor signaling between pericytes and endothelial cells promotes endothelial cadherin expression and stimulates intercellular adhesion. Pericytes also produce ECM. Astrocytes produce SHH that interacts with the Patched receptor on endothelial cells.

Obermeier *et al.* (2013) reported that when pericytes are in place they secrete ANG1 (ANGPT1 angiopoietin a potent mediator of vessel development) that interacts with a receptor TEC in endothelial cells to further promote intercellular adhesion and limit vessel permeability. This interaction enhances tight junction protein expression. WNT ligand produced by astrocytes and possibly by other progenitor cells, interacts with the Frizzled (Fzd) receptor on endothelial cells and this enhances tight junction formation.

APOE (apolipoprotein E) produced by astrocytes signals the LRP1 low density lipoprotein receptors on cerebral microvessels. The signaling activates PKC protein kinase C activity that is necessary for the post-translational modification of the tight junction proteins. There is evidence that the APOE4 variant promotes BBB disruption in that APOE4 is less efficient in the activation of PKC.

Tight junction components

These include alpha, beta and gamma catenins and cadherins. Additional components include occludins, claudins, cingulins and the zona occludens junctional adhesion molecules. The zona occludens proteins ZO1, 2 and 3) are intracellular scaffold proteins that link tight junction claudins and occludins to intra-cellular actin and the cytoskeleton.

Within the perivascular extracellular matrix foot processes of astrocytes surround capillaries. To further prevent paracellular molecular diffusion, ion channels and specific transporters in the epithelial membranes play roles in the passage of critical components for neuronal function.

Abbott *et al*. (2010) reported that large numbers of solute carriers (SLCs) occur in the BBB. SLC forms ensure transport of nutrients and organic compounds. SLCO forms transport cations and anions to the brain. The ABC transporters act as pumps that utilize ATP to transport lipid soluble molecules across the epithelial cell membrane. Macromoleules are also often transported through receptor mediated endocytosis, Abbott *et al*. (2010) noted that cell movement across the BBB is minimal. However, mononuclear cells, leucocytes, monocytes and macrophages are recruited under pathological conditions.

The paravascular pathway of cerebro-spinal fluid flow

Iliff *et al*. (2012) utilized 2-photon imaging of small molecular tracers to monitor flow of CSF into the brain parenchyma. CSF is produced in the ventricular cavities from the choroid plexus. It enters the sub-arachnoid space. Iliff *et al*. (2012) determined that there is a para-arterial influx of sub-arachnoid CSF into the brain interstitium. They further established that interstitial fluid is then cleared in an efflux pathway that runs alongside large caliber veins. The key aspect of this fluid movement is trans-astro-cytic water flow. The protein aquaporin AQP4 plays an important role in this astroglial water flux. Passage of CSF from the sub-arachnoid and peri-arterial space and its subsequent clearance through the peri-venous space is facilitated by AQP4. The perivascular end feet of astroglial cells are sites of aquaporin water channels. In addition to clearance of solutes, there is evidence that specific proteins including soluble forms of beta amyloid are cleared through this pathway.

Transcriptional Landscape of the Prenatal Brain

Miller *et al.* (2014) described results of their studies on human fetal brains at midgestation. They used magnetic resonance imaging of brains, dissection of defined regions, cyto-architectural analyses and *in situ* hybridization to study transcription. They noted that many studies of brain development have focused on mouse brain. However, there are significant differences between mouse and human brain, particularly in the neocortex. Miller *et al.* (2014) reported that the differences are due to the greater expansion of the progenitor cell pool in human. There is expansion and development of additional layers in the sub-ventricular zone. They noted that there are also apparent species differences in the origin of GABAergic interneurons. A key species difference not yet fully analyzed is the cortical evolution that underlies language development in humans.

In addition to analyzing transcription of protein coding genes, Miller *et al.* (2014) analyzed expression conserved non-protein coding sequences. They determined that these are frequently located near genes that undergo developmental regulation. They determined that in primates where rostral genes (frontal cortex) manifest increased expression, these genes are enriched for conserved non-protein coding sequences relative to caudally expressed genes. They noted that this is consistent with the expansion of the frontal cortex in primates.

Experience Driven Neuronal Activity and Gene Expression

The UBE3A ubiquitin ligase gene is expressed from the maternally inherited gene. It is expressed in the cerebellum hippocampus and neocortex. UBE3A catalyzes the addition of ubiquitin to protein targets that are subsequently degraded. There is evidence that loss of UBE3A expression impacts the nervous system primarily (Jiang *et al.*, 1998; Yashiro, 2009).

Greer *et al.* (2010) determined that in the absence of UBE3A expression excessive quantities of ARC (activity regulated cytoskeletal associated protein) accumulate in synapses and that this accumulation impairs synaptic function.

Greer *et al.* (2010) reviewed processes through which sensory, motor and cognitive stimuli shape neuronal activity. These experiences stimulate neurotransmitter release and signaling in the post-synaptic neuron and they also stimulate a rise in intra-cellular calcium and receptor insertion in the post-synaptic membrane. They emphasized that mutations in a number of components of the activity driven transcriptional program lead to autistic disorders and epilepsy.

Studies of Greer *et al.* (2010) established that synaptic depolarization of glutamate receptors leads to increased Ube3A expression. Furthermore, increased environmental stimuli induced by placing mice in an enriched environment increased Ube3A mRNA and protein. Ube3a function controls the levels of Arc protein. Since Arc protein mediates endocytosis of the AMPA receptors, Ube3A expression reduces endocytosis of AMPA receptors. Other proteins at the synapse found to be regulated through environmental stimulation included proteins known to be defective in neurobehavioral disorders or epilepsy. These genes included SLC9A6 (a solute carried defective in forms of X linked mental retardation), RSK2 RPS6KA3 (ribosomal S6 kinase defective in some cases Coffin Lowry syndrome), MECP2 (methyl CpG binding protein defective in Rett syndrome, CBP (CREB binding protein defective in Rubinstein–Taybi syndrome) and PCDH10 (defective in some cases of autism).

Greer *et al.* (2010) noted that their findings indicate that drugs that increase expression of AMPA receptors at synapses may alleviate symptoms of Angelman syndrome. They proposed that decreased levels of UBE3A likely inhibit degradation of a number of different proteins including Ephexin 5 (a neuronal guanine nucleotide exchange factor) and molecular chaperone Sacsin 5 that is highly expressed in the central nervous system.

Brain Environment Interactions: Neuroendocrine Systems

Consideration of energy balance and neuronal circuits are particularly important in development since there is growing evidence that maternal metabolic homeostasis exerts long-term effects on offspring. Clausen *et al.* (2008) reported that a hyperglycemic intra-uterine environment in human pregnancy is involved in the later pathogenesis of type 2 diabetes in offspring. Deierlein *et al.* (2011) reported that fetal exposure to high maternal glucose concentration contributed to overweight and obesity in offspring. Plagemann (2011) reported that epidemiological, clinical and experimental data indicated that increased levels of glucose and insulin in prenatal life might epigenetically program offspring to obesity and diabetes. Taylor *et al.* (2014) reported evidence that the milieu of maternal obesity in pregnancy increased in offspring cardio-metabolic risk factors, including elevated blood pressure. They proposed that maternal obesity impacted central regulatory pathways involved in blood pressure regulation.

Vogt *et al.* (2014) in their studies on mice determined that high fat diet in mothers during lactation predisposed to impaired glucose homeostasis and obesity in offspring. These were associated with impaired function of the hypothalamic and melanocortin circuitry.

There is growing evidence that neurons in the hypothalamus are crucial for regulation of energy balance and glucose homeostasis (Grayson *et al.*, 2012). Neurocircuitry in the arcuate nucleus of the hypothalamus is particularly important in both physiological processes. Circulating and local factors are sensed and responses are coordinated. Neuronal circuitries in the ventro-medial hypothalamus (VMH) and in the hindbrain are important as are neuronal connection between the intestine and the CNS.

Grayson *et al.* (2012) noted that the CNS receives signals regarding energy through hormones insulin and leptin. These signals are processed in brain nuclei in the melanocortin system. Arcuate nucleus neurons express precursor peptide POMC (proopiomelanocortin) that is processed to melanocortins, corticotropins, lipotropins and endorphins. Melanocortin is an agonist for melanocortin receptors. Another cell population in the Arcuate nucleus expresses NPY (neuropeptide Y), and AGRP a neuropeptide that is an antagonist of MC4R and MCH3R activity. Grayson *et al.*

(2012) emphasized that there is continuous cross talk between the two cell populations in the arcuate nucleus and activity of the NPY-AGRP neurons may suppress the POMC neurons. Insulin receptors and leptin receptors are present throughout the hypothalamus. They reported that the catabolic actions of insulin and leptin are in part mediated through increased expression of POMC and decreased NPY-AGPR activity. These neurons are also impacted by glucose amino acids and fatty acids.

Melanocortin System

The melanocortin system is composed of five melanocortin receptors and melanocortin alpha, beta and gamma that are peptides derived from POMC. This system includes different regions in the hypothalamus, including the arcuate nucleus, paraventricular nucleus, lateral hypothalamic area, the dorsomedial nucleus and the ventromedial nucleus. The arcuate nucleus has a semi-permeable BBB and can sense hormonal and nutrient level fluctuations. A sub-population of Gaba-ergic neurons from the arcuate nucleus interacts with other hypothalamic and hindbrain pathways to regulate energy balance.

Studies by Vogt *et al.* (2014) suggest that hyperinsulinemia during periods in fetal life or early post-natal life when melanocortin projections are being formed, may contribute to long-term impairment of the hypothalamic regulation of energy and glucose metabolism.

Gene environment interactions and neuro-cognitive and neuro-behavioral disorders

Risch *et al.* (2014) studied familial recurrence risks for autism using data from 7,559 siblings of an autism proband and data from 29,384 siblings of non-autistic controls. Results of their studies provided evidence for genetic and environmental factors in autism etiology. The overall autism risk in siblings of a child with autism was 10.1%. The autism risk in siblings of controlled children without autism was 0.52%. Of particular interest was the fact that in the next born (second born) sibling of a child with autism the autism recurrence risk was 11.5% while in later born sibling the recurrence risk was 7.3%.

In maternal half-siblings of a child with autism the recurrence risk in the next (second born) sibling was 6.5% and in later born siblings it was 3.0%. For paternal half-siblings the recurrence risk was 2.3%.

Of particular interest was the fact that shorter intervals between birth of the autism child and the next-born increased the recurrence risk of autism to 14.4%. In cases of birth intervals of 4 years or more the recurrence risk was 6.8%.

Possible prenatal causes for the effect of the shortened interval between births could be related to maternal nutritional factors e.g., folate depletion, or infections. Also shorter intervals between births are more likely to be associated with low birth weight and pre-term birth. Risch *et al.* (2014) noted that there are numerous reports in the literature relating shorter inter-birth intervals with a higher frequency of neurodevelopmental problems. Possible post-natal factors that may be associated with shorter birth intervals include problems related to breast feeding and coping problems.

Neurogenesis in the Adult Brain

Adult neurogenesis occurs in the sub-granular zone (SGZ) of the dentate gyrus and in the sub-ventricular zone of the lateral ventricles. Ma *et al.* (2010) reviewed adult neurogenesis. Adult neural stem cells in the SGZ of the dentate gyrus in the hippocampus proliferate mature and are integrated into neural circuitry and contribute to learning and memory. Neural stem cells in the sub-ventricular zone generate glia and neuroblasts. The latter migrate to the olfactory bulb.

Ma *et al.* (2010) emphasized that epigenetic mechanisms play key roles in the regulation of adult neurogenesis and interface with environment, experience and physiological state and with cell signaling and regulatory mechanisms. The key epigenetic mechanisms include DNA methylation, histone modification, chromatin remodeling and transcriptional feedback loops. The latter involve transcription regulators and non-coding RNAs. Factors that regulate DNA include generation of 5-methylcytosine and DNA methylation and repair of DNA through base excisions repair (BER) and nucleotide excision repair (NER). Histone modifications require lysine methyltransferase, lysine acetyltransferase and arginine methyltransferases.

Ma *et al.* (2010) noted that epigenetic processes can lead to long-term changes and can also play roles in timing and in transition from one cellular state to another. Analysis of sequential development and cell state transitions are dependent upon availability of cell specific antigens that facilitate cell isolation. In addition, sensitive bisulfite sequencing methods and chromatin immunoprecipitation and sequencing methods are necessary to analyze DNA and chromatin modification.

Ma *et al.* (2010) determined that in the sub-granular zone neural cells that are positive for nestin and GFAP differentiate into neural progenitors that are positive for TRIB2 (tribbles a signaling pathway modulator and then into DCX, PSA (PSAT1 a pyridoxine dependent aminotransferase) and neural cell adhesion molecule (NCAM) positive neurons, then into new neurons. Laser capture permits isolation of homogeneous populations of various cell types. Epigenetic profiling of distinct cell populations can be carried out. They demonstrated that methyl CpG binding protein MBD1 plays a key role in the regulation of adult SGZ neurogenesis.

Neuronal stem cell differentiation is regulated in part by the Trithorax group of proteins that facilitate maintenance of H3K4 methylation. MLL1(KMT2A) is a member of the TRX group of proteins and acts as H3K4 methyltransferase required for neuronal stem cell proliferation. The homeobox gene DCX is targeted and methylated by MLL2 (the KMT2D methyltransferase). The JMJD3 (Jumonji domain protein KDM6B) is also important for neural stem cell differentiation; it acts as a histone3 lysine 27 (H3K27) demethylase.

Gene activation is also influenced by histone acetylation carried out by histone acetyltransferase (HAT). Gene expression is also impacted through histone deacetylases. Ma *et al.* (2010) reported that histone deacetylases silences expression of NeuroD neuronal differentiation transcription factor. HDAC silencing is apparently less stable and more dynamic than other forms of silencing. Non-coding RNAs including micro RNAs also play roles in regulating differentiation.

Long interspersed repeats line elements and neuronal diversity

LINE (L1) elements are retrotransposons that constitute 17% of the human genome and these elements are silenced by methylation. Muotri *et al.* (2005) and Coufal *et al.* (2009) determined that L1 elements can be activated and undergo retrotransposition during neuronal differentiation and that this leads to modification of the genome. Genome modifications in specific neurons play roles in creating neuronal diversity and somatic mosaicism occurs in the adult brain. Methylation of L1 elements is reported to be less than in other tissues.

Environmental stimuli and niches in adult neurogenesis

Neurogenesis in the adult brain is impacted by environmental stimuli and by local signals from the niche surrounding the neural stem cells. Expression of GADD45B in neurons near the SGZ leads to demethylation of promoters of specific genes and to active expression of brain derived neurotrophic factors (BDNF) and FGF2 (fibroblast growth factor a mitogen for neural stem cell proliferation). Ma *et al.* (2010) proposed that GADD45B functions as a sensor for environmental stimuli. They noted that endothelial cells and glial cells in the SGZ and the SVZ act as niches for adult neurogenesis.

Faigle and Song (2013) reviewed factors that impact adult neurogenesis. They reported that specific signaling pathways play roles in adult neurogenesis, neural stem cell proliferation and integration of newly developed neurons into brain circuitry.

The WNT signaling pathway is involved in several stages of nervous system development from the neural tube stage onward. There is also evidence that WNT signaling plays an important role in adult hippocampal dentate gyrus neurogenesis.

NEUROD1 is a proneurogenic transcription factor that is induced by WNT signaling. When WNT is present beta catenin moves to the nucleus and participates in NEUROD1 expression. There is evidence that the WNT/beta catenin pathway promotes proliferation rather than differentiation of adult neural stem cells. Faigle and Song (2013) reported that there is evidence that WNT signaling also plays roles in adult neurogenesis in the sub-ventricular zone.

Notch signaling pathway activity is involved in neural stem cell maintenance and in the sub-granular zone it is also involved in the proliferation of stem cells. The SHH pathway plays roles in stem cell progenitor proliferation and in neural stem cell maintenance and in neuroblast migration.

The growth factor BDNF increases neurogenesis. Other growth factors that enhance neurogenesis include fibroblast growth factors (FGF), insulin like growth factor (IGF2) and neurotrophin 3 NTF3

Studies by Ernst *et al.* (2014) in humans revealed that neuronal precursor cells (neuroblasts) generated in the sub-ventricular zone were integrated into the striatum. They demonstrated through transcriptome analysis that DCX, a neuroblast marker is expressed in human hippocampus and striatum. In addition another neuroblasts markers polysialylated neural cell adhesion marker (NCAM) was shown to be expressed in hippocampus, striatum and cerebellum.

Ernst *et al.* (2014) also carried out retrospective dating of neuronal cell birth based on their C14 content. This birth dating is based on the fact that C14 released during atomic testing during the Cold War was incorporated into proliferating cells. Using measurements in subjects aged 3 to 79 years along with histological and histochemical cell analysis, Spalding *et al.* (2013) determined that neurogenesis occurred in post-natal and adult life. Using the C14 birth dating Ernst *et al.* (2014) determined that postnatal and adult neuroblasts were incorporated into the hippocampus, dentate gyrus and striatum and these neuroblasts give rise primarily to interneurons. They noted that in the striatum of patients with Huntington's disease there are few neuroblasts in the striatum.

Ernst *et al.* (2014) raised the possibility that the fact that neuroblasts are continually born and make their way to the striatum could be exploited for therapeutic purposes. The question that arises then is whether specific factors or processes could promote generation or survival of new neurons.

CHAPTER 8

CRANIO–FACIAL DEVELOPMENT AND DEFECTS

Introduction

Twenty-two different bones comprise the skull and face and in addition there are deciduous and secondary teeth. Wilkie and Morriss-Kay (2001) reviewed aspects of cranio–facial development. They defined the neurocranium and the viscerocranium. The neurocranium includes frontal, parietal, sphenoid and temporal bones. The viscerocranium includes mandible, maxilla, zygoma and nasal bones. The neural crest is important in the formation of the viscerocranium and in formation of part of the neurocranium.

Holoprosencephaly

Wilkie and Morriss-Kay (2001) also reviewed several of the major malformations of the face and skull. In holoprosencephaly malformations of the brain, skull and midline malformation of the upper face occur. The degree of severity of the malformation varies in different individuals. Mid-line abnormalities develop in part due to obstruction of migration of fronto-nasal neural crest cells.

Heterozygous mutation in any one of at least four different genes can lead to holoprosencephaly. These genes include *SHH* sonic hedgehog, *SIX3* homeobox gene, *ZIC3* (zinc finger transcription factor) and *TGIF1* (transcription regulator). Defects in the cholesterol synthesis pathway can lead to holoprosencephaly, in part because SHH protein requires post-translation modification with cholesterol related molecules.

Wilkie and Morriss-Kay (2001) emphasized the important roles of fibroblast growth factor receptors (FGFRs) in development of the skull vault. Defects in the function of these receptors lead to disorders characterized by abnormal skull shape and abnormal fusion of skull sutures (craniosynostosis), e.g., Apert, Pfeiffer and Bears–Stevens syndromes. A different skull vault malformation syndrome is caused by heterozygous loss of function mutations in the TWIST protein. This protein normally acts to suppress histone acetyltransferase activity and gene transcription.

Abnormal Head Size

Abnormally large head size, macrocephaly, and abnormally small head size, microcephaly occur in the context of a number of congenital malformation syndromes. Macrocephaly occurs and in the overgrowth syndromes, Soto's syndrome, Soto's like syndrome, Weaver syndrome and Beckwith Wiedemann syndrome (discussed in section on growth abnormalities). Deletions or duplications in the chromosome 1q21.1 region were reported to be associated with abnormal head size (Brunetti-Pieri *et al.*, 2008).

Craniosynostosis

In this condition, there is premature fusion of one or more of the skull sutures. In some cases craniosynostosis occurs as part of a malformation syndrome. However, Lattanzi *et al.* (2012) reported that in over 85% of cases, craniosynostosis occurs in the absence of malformation in other regions. Premature fusion of one or more of the cranial sutures leads to abnormal head shape.

Molecular genetic studies in a subset of patients with craniosynostosis have led to identification of mutations in a number of genes associated with intra-membranous ossification. These genes include the FGFRs: FGFR1, FGFR2 and FGFR3; also insulin like growth factor receptor IGF1R, transcription factors TWIST and RUNX2. Lattanzi *et al.* (2012) reported that TWIST1 is a transcriptional regulator of mesenchymal cells during skeletal development and that it plays roles in both chondrogenic and osteogenic differentiation. IGF1R is a receptor for the insulin like growth factor IGF1. IGF1 and IGF2 growth factors are implicated in osteogenesis in sutures. The *RUNX2* gene encodes a pro-osteogenic which is also expressed in sutures. Mefford *et al.* (2010) reported copy number variants (duplications) of the *RUNX1* gene in cases of craniosynostosis. Copy number variants of the 1p36.3 chromosome regions have also been reported in cases of craniosynostosis.

Neural Crest

Simoes-Costa *et al.* (2014) reviewed information on neural crest cell region of origin and their migration. The region in the early embryo that gives rise to the neural crest is located bilaterally in regions between the neural plate and the adjacent ectoderm. With neurulation and invagination of the neural plate and subsequent closure of the neural tube, the border regions from both sides containing the neural crest cells are elevated and they then fuse. Subsequently the intercellular connections between the cells are lost (delamination), cells become separated and they undergo epithelial to mesenchymal transition and then begin to migrate to different sites in the embryo. There are four major regions of neural crest cell migration: cranial, vagal, trunk and sacral. Neural crest cells in the different regions then undergo site-specific differentiation. In the cranial region, neural crest cells contribute to formation of the cranio–facial skeleton and cranial nerve ganglia. The vagal population of neural crest cells contributes to smooth muscle of blood vessels and heart and to enteric ganglia. The trunk neural crest cells contribute to the dorsal root ganglia, sympathetic nerves and adrenal glands. The sacral population of neural crest cells also contributes to enteric ganglia. Neural crest cells contribute to neurons glia and Schwann cells in the peripheral nervous system. Neural cells in all regions give rise to melanocytes.

Takahashi *et al.* (2013) emphasized that the migration pathways and site-specific differentiation patterns are impacted by input of external ligands that bind to receptors on the neural crest cells. Simoes-Costa *et al.* (2014) emphasized the importance of the neural crest gene regulatory network in neural crest cell development. This network includes transcription factors and signaling molecules. Formation of the neural crest border is dependent upon signaling pathways that include WNT, FGF, BMP (bone morphogenetic proteins) and NOTCH, and expression of transcription factors, including DELTA and homeobox factors MSX, PAX3, PAX7, DLX3 and DLX5. Delamination and the subsequent epithelial to mesenchymal transition is dependent upon transcription factors SNA1 and 2, TFAP2A, FOXD3, TWIST, SOX 9 and 10. The transcription factor binding protein ID1, and the phosphoprotein encoded by the *MYC* gene also play roles. As neural crest cells activate expression of receptors, they are able to respond to environmental signals. Chromatin modifications play important roles in regulating this gene expression.

Neural Crest Cells and Cranio–Facial Developmental Defects

Cleft lip, cleft palate

Neural crest cells and mesenchymal cells from the first and second branchial arches participate in the development of the face.

Oral clefts are among the most frequently occurring congenital malformations. The incidence is approximately 2 per 1000 births. However, there are differences in the population frequencies of oral clefts.

Dixon *et al.* (2011) in reviewing cleft lip and palate reported that by the sixth week post-conception in the human embryo, medial processes merge with the maxillary process to form the upper lip and primary palate. During the sixth week bilateral out growths of the maxillary processes grow vertically down and subsequently assume a position above the tongue and then fuse with each other, giving rise to the secondary palate. These different developmental steps explain differences in the types of oral clefts seen at birth.

There is abundant evidence that genes and environmental factors play roles in the etiology of oral clefts. Syndromic and non-syndromic forms occur. In syndromic forms, oral clefts occur along with other congenital malformations, while in the non-syndromic form the oral cleft occurs as a single congenital malformation. Dixon *et al.* (2011) reported that 70% of cases where cleft lip and palate occur are non-syndromic and 50% of cases of cleft palate only are non-syndromic.

Insights into genetic factors involved in oral clefting were obtained through discovery of the underlying gene defects in the syndromic form of oral cleft, van der Woude syndrome.

Van der Woude syndrome is a rare autosomal dominant disorder with variable expression. Manifestations of this disorder include cleft lip and palate, congenital lip pits and mounds, and sometimes teeth abnormalities (hypodontia). This autosomal dominant condition was mapped to human chromosome 1 through linkage studies. Intense subsequent efforts led to identification of the gene *IRF6* that harbored causative defects. *IRF6* encodes interferon regulatory factor. Studies in mouse models of the mutant *IRF6* gene revealed that mutations are associated with hyperproliferative epidermis in the oral cavity and that this epidermis fails to undergo terminal differentiation. Multiple epithelial adhesions result from differentiation failure and these adhesions promote clefting.

A specific enhancer element upstream of *IRF6* was identified and a specific variant in this region was shown to be over-transmitted from parents to offspring affected with non-syndromic cleft lip. Furthermore, this enhancer was shown to bind a specific transcription factor TFAP2A. This transcription factor was previously found mutated in a syndromic form of cleft lip and palate (Dixon *et al.*, 2011).

Ferrero *et al.* (2010) described a family with a novel *IRF6* mutation and evidence of highly variable expression of van der Woude syndrome. Some family members had only lip pits and lip mounds while the youngest member of the family had cleft lip and palate and lip pits.

A locus on chromosome 8 is linked to non-syndromic cleft lip and palate with high significance (Mangold *et al.*, 2011).

Multifactorial threshold model for cleft lip and palate

It is important to note that analysis of family data and inheritance patterns of non-syndromic oral clefting provide evidence for a multifactorial threshold model of inheritance. Grosen *et al.* (2010a) studied recurrence of oral clefting in 54,000 relatives of oral cleft cases in Denmark. Recurrence risks in first degree relatives ranged between 3.2 and 4.7% and recurrence risk in offspring ranged between 4.1 to 5.1%. In second-degree relatives, half-siblings, nieces and nephews, the recurrence risk ranged between 0.8 and 1%. In third degree relatives, cousins, the relative risk ranged between 0.6 and 0.8%. Severe forms of clefting impacted the degree of inheritance to some degree. The background population risk for oral clefting was 0.18.

Grosen *et al.* (2010b) also analyzed recurrence risk for oral clefting in offspring of twin pairs discordant for oral clefts. They determined that in children of 54 affected twins, the recurrence risk was 1.8%. In children of 65 unaffected twins, the risk of clefts was approximately 2.3%. They concluded that the recurrence risk was similar in the offspring of unaffected and affected twins. These studies support the role of genetic factors and likely multi-factorial inheritance patterns.

Environmental Factors in Oro-Facial Clefting

In a population based study of infants and fetuses born with oro-facial cleft deformities in California between 1999–2003, Wallenstein *et al.* (2013) reported an at least two fold risk of cleft palate in women with low intake of riboflavin, vitamin B12, calcium, zinc and magnesium. For cleft lip and palate they found a two-fold elevated risk with low intake of niacin, riboflavin, vitamin B12 and calcium and a decreased risk with high intake of folate.

Mandibulo-Facial Dysostoses

Mandibulo-facial dysostoses are characterized by an abnormally small mandible–small zygomatic arches, small ears, hearing loss, down-slanting palpebral fissures and coloboma (growths) of the eyelid. Treacher–Collins syndrome is an example of mandibulo-facial dysostosis, other forms also occur. In addition, there are disorders where facial dysostosis is accompanied by other skeletal abnormalities that involve limbs. These disorders are sometimes referred to as acro-facial dysostosis (Wieczorek, 2013).

Treacher–Collins syndrome is primarily due to mutations in the TCOF1 gene leading to deficiency in the protein that it encodes, TREACLE. This protein plays role in ribosomal maturation and insufficiency of the protein impairs ribosomal function and leads to impaired proliferation of neural crest cells.

Two additional Treacher–Collins related syndromes are due to defects in the RNA polymerases POLRID and POLRIC that both play roles in ribosomal RNA transcription.

Several gene defects that lead to different forms of acro-facial dysostosis have also been described. Nager syndrome is a mandibulo-facial dysostosis plus abnormalities of the thumbs and fingers. Mutations in the gene *SF3B4* have been found in this syndrome. SF3B4 encodes a component of the splicing factor 3B and also apparently impacts osteochondral cell differentiation. Another spliceosomal component encoding gene *EFTUD2* was shown to be defective in Gulon Almeida acrofacial dysostosis associated with mandibulo-facial dysostosis features and in addition microcephaly and cleft-lip and palate may be present. EFTUD2 protein is a GTPase that is a component of the spliceosome complex.

Waardenburg Syndrome

Defects in any one of six genes that are involved in neural crest cell differentiation to melanocytes and melanocyte proliferation can lead to Waardenburg syndrome. Features of this syndrome include depigmented patches of hair and skin, heterochromia of the eyes, sensorineural hearing loss. The peripheral nervous system, enteric nervous system and muscles may be involved in some cases. Pingault *et al.* (2010) reviewed the different clinical forms of Waardenburg syndrome and underlying gene mutations. Heterozygous mutations or deletions of the *PAX3* gene occur in type 1 Waardenburg syndrome (WS1). PAX3 is a transcription factor involved in the differentiation of neural crest derivatives including melanoblasts. The *MITF* gene (microphthalmia associated transcription factor) regulates melanocyte development. Pingault *et al.* (2010) reported that heterozygous mutations in this gene lead to approximately 15% of cases of WS2. Other genes that lead to WS2, 3 or 4 when mutated include endothelin gene *EDN3*, endothelin receptor EDRNB and transcription factor encoding genes *SOX10* and *SNAI2*. The latter encodes a zinc finger transcription factor. SOX10 and SNAI2 transcription factors are involved in the early stages of neural crest cell differentiation.

Tumors that originate in cells of neural crest origin and that occur in infants are described in the section on abnormal growth.

CHAPTER 9

MOLECULAR ASPECTS OF HEART DEVELOPMENT

Stages in Heart Development

Information on the early stages of heart development is derived primarily from studies of heart development in vertebrate model organisms, from studies on patients with congenital heart disease and more recently from studies of cardiac cell differentiation in cultures of embryonic stem cells or of induced pluripotent stem cells.

The earliest stage of cardiac development in vertebrates involves induction of mesenchymal cells in the primitive streak on the embryonic plate to move laterally from the primitive streak to form left and right heart forming regions. Cells from these regions then migrate cranially and coalesce in the midline to form a crescent. The cardiac crescent is sometimes referred to as the primary heart field (Wagner and Siddiqui, 2007). Specific growth factors control induction and movement of cardiac progenitor cells. These growth factors include bone morphogenetic proteins (BMP2), fibroblast growth factors (FGF8) and receptors (FGFR1). Negative factors limit the extent of migration of cells of the primary heart field. These negative factors include members of the WNT signaling pathway and NOGGIN [NOG: a secreted protein that inactivates transforming growth factor beta (TGFbeta) signaling proteins]. Following migration of cells, expression of transcription factors within cardiogenic mesenchymal cells is activated (Mummery *et al.*, 2012). These transcription factors include T (T Brachyury homolog: a nuclear transcription factor), MIXL1 (homeodomain protein), MESP1 (mesoderm helix loop helix protein), ISL1 (transcription factor that binds to enhancers of several genes including insulin gene) and TBX5 (T box transcription factor).

A secondary heart field develops adjacent to the primary field. There is evidence that this secondary field ultimately contributes to the outflow tract, the conus and truncus. Wagner and Siddiqui (2007) noted that FGF8, BMP2 and Sonic hedgehog (SHH) were most important in the development of this region. These signaling stimuli apparently direct expression of transcription factors TBX1, TBX5, GATA4 and NKX2-5.

The cells in the cardiac crescent region fold to form the heart tube and cells from the secondary heart field migrate to this. The next step in development involves looping of the tube, chamber formation, and establishment of left to right asymmetry in the embryo. The TGFbeta related molecules LEFTY (left right determination) and NODAL play key roles in these processes. Hedgehog (HH) pathway signaling plays roles in looping and left–right axis establishment.

Wagner and Siddiqui (2007) reported that key events in the functional specification of ventricular chambers include formation of trabeculated myocardium along the outer curvature of the heart tube. Major signal transduction pathways involved in formation of trabeculated myocardium include the retinoic acid pathway and the neuregulin/ErbB pathway. Specific levels of expression of BMP10 are also necessary for optimal proliferation of cardiomyocytes.

Key events in determination of direction of blood flow are the development of heart valves. Wagner and Siddiqui (2007) reviewed the development of endocardial cushions, the primordia of the heart valves. The extra-cellular matrix between the myocardial layer and the inner endocardial layer is referred to as cardiac jelly. At specific location in the atrio-ventricular canal the cardiac jelly swells and cells migrate into it. These cells that form the endocardial cushions have undergone epithelial mesenchymal transformation. There is evidence that several different signals sequentially expressed determine endocardial cushion formation, they include vascular endothelial growth factor (VEGF); also important are epidermal growth factor (EGF) related signals. Progression of differentiation from endocardial cushions to heart valve leaflets also involves VEGF and the nuclear transcription factor NFAT.

Mummery *et al.* (2012) emphasized that the sequential activation of expression of specific transcription factors is key to appropriate differentiation. They noted that expression of specific cardiac structural proteins was also critical; these proteins include actin, myosin, troponins, desmin and elastin.

Congenital Heart Disease in Humans

Congenital heart disease (CHD) affects 1% of the population (Andersen *et al.*, 2014). Many different genomic and genetic changes can lead to this disease. Furthermore, CHD can occur as part of a syndrome or can occur as an isolated malformation.

Genomic changes in congenital heart disease

Chromosomal aneuploidies, e.g., trisomies of chromosome 21, or chromosome 18 or 13, and XO are major contributors to etiology of CHD. In trisomy 21, 40–50% of cases have CHD and in XO individuals (Turner syndrome) 20–50% of individuals have CHD. Fahed *et al.* (2013) noted that although any type of heart defect can occur in these cases of chromosome aneuploidy, there are specific lesions that occur more commonly in a specific aneuploidy. For example, atrio-ventricular septal defects occur frequently in trisomy 21 and co-arctation of the aorta occurs in some cases of in Turner syndrome (XO).

Specific segmental chromosome defects (copy number variants, CNVs) leading to dosage changes, are also frequently associated with CHD. These segmental defects often encompass a number of genes. Defining which specific gene within the altered segment leads to CHD has been possible through follow-up studies in series of cases and also through knowledge of specific genes that lead to heart disease in model organisms.

Segmental chromosome changes associated with congenital heart disease include 22q11.2 deletions that include deletion of TBX1, 7q11.2 deletions that delete elastin ELN and chromatin modifier BAZ1B and lead to Williams syndrome and deletions on 1p36 that delete DVL (disheveled a gene in the WNT signaling pathway).

There is some evidence from studies on large cohorts with congenital heart disease that the frequency of small CNVs is higher in affected individuals than in controls. The CNV number is highest in cases with congenital heart disease and other malformations than in cases with isolated CHD (Anderson *et al.*, 2014).

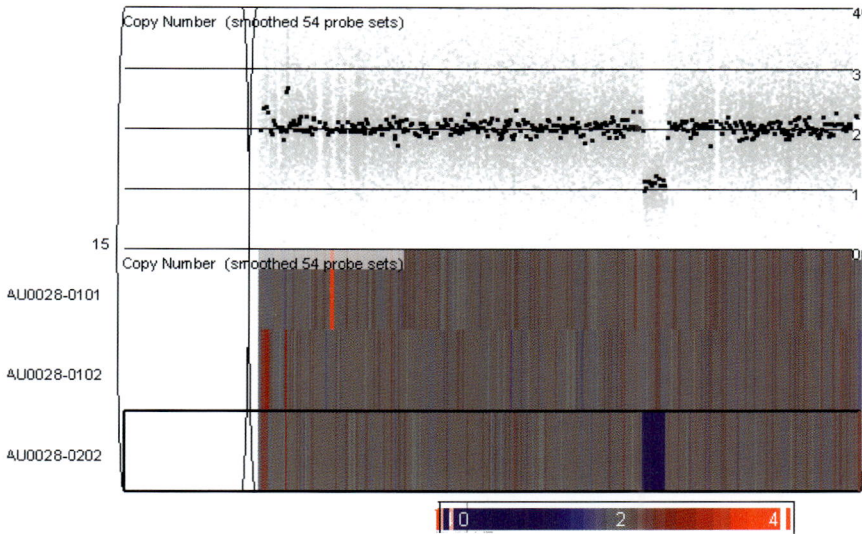

Fig. 9.1:   Microarray showing deletion 15q24.

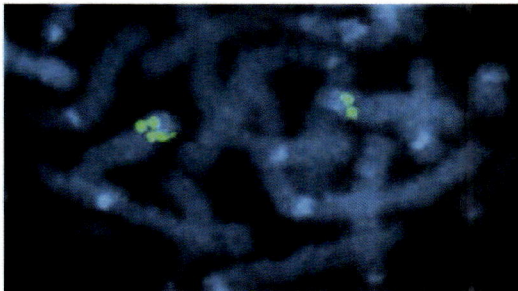

Interstitial duplication of 15q12-q13
demonstrated with Bac clone RP11-41H23

Fig. 9.2:   Interstitial duplication on chromosome 15q12–q13.

Gene mutations that cause isolated congenital heart disease

Fahed *et al.* (2013) presented data on genes found to be defective in cases where CHD was the only developmental abnormality. These included 15 transcription factors or co-factors, 17 receptors, ligands or downstream effectors, five structural proteins including myosins, elastin and actin. There is also evidence that mutations in genes that are involved in chromatin modification lead to CHD in some cases.

Zaidi *et al.* (2013) reported that in the CHD cases there was an excess of *de novo* mutations in genes that encode products involved in reading or activating modification of histone 3 lysine 4 (H3K4), e.g., MLL2 (KMT2D) that acts as a lysine specific methyltransferase and demethylases involved in inactivation of H3K27 (histone 3 lysine 27). Other genes that impact chromatin also showed mutations. These included SMAD2 that enters the nucleus and binds to chromatin, and CHD7 that acts as a chromodomain helicase.

MLL2 (KMT2D) mutations occur in some patients with Kabuki syndrome, however, the CHD patient with MLL2 mutation in the Zaidi study did not have Kabuki syndrome. Similarly the CHD patient with the CHD7 mutation in their study did not have CHARGE syndrome, a syndrome that is associated with CHD7 mutations.

Fahed *et al.* (2013) reported data on dominant mutations (i.e., heterozygous mutations) in transcription factors that caused heart disease. Mutations in CHD patients occurred in NKX2.5, NKX2.6, GATA4, GATA5, GATA6, IRX4 (Iroquois homeobox), TBX20, and ZIC3 zinc finger protein) and in transcription co-factor FOG2 (friend of GATA2). Key signaling pathway genes found to be mutated in CHD included NODAL, LEFTY, NOTCH, JAG1 (Jagged 1 is a ligand for NOTCH).

Systems biology approaches and construction of molecular networks in heart development have revealed additional possible candidates as CHD causing genes.

Fayed *et al.* (2013) emphasized an additional hypothesis namely that altered levels of a number of different developmentally important gene products could act together to lead to heart disease. Furthermore, modifier genes could lessen the phenotypic effects of a specific mutation.

Insight into CHD causing gene mutations through next generation sequencing

Zaidi *et al.* (2013) carried out exome sequencing of CHD affected children and their parents (trios). They analyzed sequence from 362 CHD trios and from 264 control trios.

They selected 363 probands with severe CHD who had no first degree relatives affected with CHD.

Analysis of data on genes with high expression in the developing heart revealed protein altering damaging changes in 54 of 362 CHD cases (0.15) and in 15 of 264 control cases (0.05). These difference between CHD cases and controls were significant (p = 0.0005).

In *de novo* cases support for disease relevance: a particular deleterious non-synonymous variant may be obtained if different unrelated cases have the same change. Further support for the CHD relevance of a specific mutation would also be obtained if the mutations occurred in a gene that is known to participate in a specific step in cardiac development. Fahed *et al.* (2013) noted that definitive CHD causing mutations identified thus far usually perturb development through haploinsufficiency and reduction in protein products.

Mutations in cases of syndromic heart disease

Alagille syndrome is characterized by skeletal abnormalities, distinct facial features cholestasis, and CHD that may include tetralogy of Fallot, pulmonary stenosis, or peripheral pulmonary hypoplasia. It arises due to mutation in the genes JAG1 that encodes a Notch ligand and in some cases it is due to mutations in NOTCH2.

Holt–Oram syndrome arises from dominant loss of functional mutation in TBX5 and is associated with atrial or ventricular septal defects, conduction system defects and limb abnormalities, particularly upper limb abnormalities.

Noonan syndrome is associated with short stature, developmental delay, webbing of the neck and cardiac abnormalities including atrial septa defects, pulmonary valve stenosis, and co-arctation of the aorta. This syndrome is caused by mutations in any one of 11 genes in the RAS, MEK pathway including PTPN11, a protein tyrosine phosphatase.

Insight into Cardiac Development through Stem Cell Studies: Induction of Cardiomyocytes from Induced Pluripotent Stem Cells

Generation of cardiomyocytes from patients with congenital or inherited diseases provide opportunities to characterize genes that lead to CHD and to study pathophysiology in specific patients.

Mummery *et al.* (2012) noted that embryonic and induced pluripotent stem cells can be rapidly expanded but that differentiation has been more difficult to achieve. They reviewed protocols to derive differentiated cardiac myocytes in monolayer cultures. The first sign of differentiation is frequently the appearance of foci of contractile cells in culture. However, additional functional tests are required to investigate differentiation to cardiac myocytes. These additional tests include studies to determine the presence of cardiac specific proteins including troponin, myosin heavy chains, and myosin light chains. Measurement of the generation of action potentials using microelectrodes is also important to establish functionality.

CHAPTER 10

VASCULOGENESIS MALFORMATIONS
AND HEMATOPOIESIS

Vasculogenesis

In early embryonic life, approximately on day 17, hemangioblastic aggregates develop in the yolk sac mesoderm. These aggregates give rise to primitive hematopoeitic cells and endothelial precursor cells. Shortly thereafter endothelial precursor cells also arise in the embryonic disc. The endothelial cells then form vasculogenic cords that differentiate into small vessel networks (Larsen, 2009).

Studies on vascular malformations have provided important insights into processes in vessel development and causes of vascular malformations. Brouillard and Vikkula (2007) reviewed genetic causes of vascular malformations. They noted that blood vessels and lymphatic vessels are formed from a single layer of endothelial cells surrounded by layers of vascular smooth muscle cells and pericytes. Pericytes are mural cells embedded in the basement membrane at the outer surface of endothelial cells. Brouillard and Vikkula (2007) sub-classified congenital vascular anomalies into vascular tumors (hemangiomas) and vascular malformations. Vascular malformations are present at birth and grow proportionally as the child grows.

Venous Malformations

These include venous mucocutaneous malformations (VMCM) and glomuvenous malformations (GVM). The venous mucocutaneous malformations usually involve skin and mucosa but may sometimes infiltrate deeper into tissue. Most often these lesions occur sporadically but in some cases they occur in several members of a family. In a number of families hereditary VMCM arise due to mutations in the TIE2 gene that encodes a tyrosine kinase receptor specific to endothelial cells. The most common mutation is R849W and this leads to an autosomal dominant form of venous malformation. However other mutations in TIE2 have been reported. The TIE2 protein acts as an angiopoietin receptor. Three different forms of the angiopoietin can bind to this receptor.

GVM

These are pink to blue purple patches that involve skin and sub-cutaneous tissue; the lesions are frequently multifocal but seldom involve mucous membranes. These lesions may be painful on palpation and contain abnormally differentiated vascular smooth muscle cells referred to as glomus cells. Glomuvenous malformations arise due to mutations in glomulin, a phosphoprotein encoded by the GLMN1 gene. The mutations are inherited in an autosomal dominant manner. Amyere *et al.* (2013) reported that somatic second hit mutations are required to trigger formation of the lesions. They noted that the second hit mutations are often gene rearrangements.

Glomulin binds to the ring domain of the RBX1 protein (ring box ubiquitin ligase) and inhibits the ubiquitin ligase activity of that protein.

Capillary Malformations

Brouillard and Vikkula (2007) noted that capillary malformations include red-purple flat lesions that frequently involve the head and neck. Some of these lesions, for example nevus flammeus, occur in newborns and fade progressively later. Large capillary lesions such as port-wine stains may however be long lasting.

Sturge Weber syndrome is characterized by portwine stain that occurs in skin enervated by the ophthalmic branch of the trigeminal nerve. Shirley *et al.* (2013) noted that in this syndrome there are also frequently venous capillary malformations of the leptomeninges and the eye. They reported that a child with a facial portwine stain has a 6% chance of having the Sturge Weber syndrome with additional associated lesion. They carried out whole genome sequencing and paired sample analysis (an affected tissue and an unaffected tissue or blood) from nine patients with Sturge Weber syndrome and reported that a mutation 548G>A in the GNAQ gene on chromosome 9q21 occurred in the portwine lesions but was negative in normal skin. In 12 of 13 patients with the syndromic form of portwine stain, they identified the same mutation. The mutant allele constitutes 1 to 18% of the DNA present in a specific lesion.

GNAQ is a guanine nucleotide binding protein. Shirley *et al.* (2013) determined that the mutation in GNAQ led to a significant increase in the activation of the extra-cellular signal regulated kinase ERK.

Shirley *et al.* (2013) noted that discovery of the somatic activating properties of the mutation in GNAQ confirmed a hypothesis of Rudolf Happle proposed in 1987 that sporadic asymmetric scattered birth defects involving the skin are likely caused by somatic mosaic mutations.

Cerebral cavernous malformations (CCM)

These malformations may be asymptomatic for long periods until they give rise to seizures and neurological manifestations; Brouillard and Vikkula (2007) described these lesions as dilated capillary like and saccular vessels with thickened walls. The endothelial cells in these lesions lack adherence and tight junctions. A number of different gene loci have been linked to the disorder. In CCM1, the KRIT1 protein is defective. This protein participates in cell-cell interactions, it associates with integrin and with microtubules. In CCM3, there are mutations in the protein PDCD10, a phosphotyrosine scaffold protein.

Telangiectasias

Brouillard and Vikkula (2007) described these malformations as focal dilatations of post-capillary vessels with excessive layers of vascular smooth muscle cells. Telangiectasias occur in a number of different genetically determined disorders. In the autosomal dominant disorder hereditary hemorrhagic telangiectasia (HHT), also known as Rendu Osler Weber syndrome, mucocutaneous telangiectasias occur along with arterio–venous malformations in organs including lung, liver and brain. Bleeding from the mucocutaneous lesions sometimes occur. Four different loci have been linked to this disorder. The HHT1 gene locus encodes endoglin. Endoglin occurs on the surface of cells and forms a complex with growth factor receptors. The HHT2 locus encodes an activin receptor kinase.

Telangiectasias are also a feature of the disorder ataxia telangiectasia due to mutation in the ATM protein that participates in the cell cycle checkpoint mechanisms.

Lymphatic Malformations and Lymphedema

Lymphatic malformations and dilations sometimes occur that are not connected to the lymphatic system (Brouillard and Vikkula, 2007).

Primary congenital lymphedema (Milroy syndrome) may be present at birth and manifest with pleural or peritoneal effusion. Later it is most commonly associated with swelling of the lower limbs. This disease is due to mutations in the vascular endothelial growth factor receptor 3 (VEGFR3, also known as FLT3). New mutations occur and the disease then follows a sporadic inheritance pattern. However mutations are sometimes inherited.

Late onset lymphedema that develops around puberty is in some cases due to mutations in the FOXC2 transcription factor.

Hematopoiesis

In humans hematopoiesis occurs initially on day 17 in extra-embryonic mesoderm in the yolk sac and gives rise to primitive hematopoietic cells (Larsen, 2009). Hematopoietic cells have to divide and self-renew to form a pool of stem cells and they have to differentiate into cells of the blood lineage. Several subsequent phases of hematopoiesis occur. Changes in site of hematopoiesis in the developing fetus occur in part as a result of the changing anatomy. In addition signals from different microenvironments play roles in supporting differentiation of stem cells (Mikkola and Orkin, 2006).

The first phase of hematopoiesis in the yolk sac gives rise to primitive erythroid cells, to macrophages and megakaryocytes. A second phase of hematopoiesis in the yolk sac gives rise to erythroid cells, megakaryocytes and myeloid lineages (Baron *et al.*, 2012). Other sites for hematopoiesis include the aorta–gonad–mesonephros mesenchyme and additional vascular sites. There is some evidence that the placenta may contribute to hematopoiesis.

Later hematopoietic cells seed the fetal liver and then multiply. It is likely that the fetal liver provides signals that facilitate hematopoietic stem cell (HSC) division. Later in gestation (after 70 days) hematopoiesis is established in the bone marrow.

Steps in hematopoietic cell differentiation

Many of the earlier studies that delineated steps of differentiation of hematopoietic cells were carried out in mice. More recently human hematopoiesis has been intensely studied and it is evident that differences exist in mice and humans in specific steps in the process. Studies of cultured human stem cells provide insight into steps in blood cell differentiation; however it is important to note that the steps in differentiation in these cultures may differ from those that occur during pre-natal life.

Doulatov *et al.* (2012) discussed the clinical importance of understanding human hematopoiesis. They noted that in bone marrow only 1 in 10^8 cells is a transplantable stem cell and purification of these cells is therefore complex. Cell sorting and purification processes based on the use of cell surface markers have historically been applied to this process. Doulatov *et al.* (2012) emphasized that understanding the events involved in the passage from HSC with self-renewal capacities to lineage committed progenitor cells is the key.

Doulatov *et al.* (2012) emphasized that multipotent HSCs reside at the apex of the hematopoietic hierarchy followed by multipotent progenitor cells (MPPs). The lymphoid progenitor cells (MLPs) give rise to the lymphoid branch, T, B and NK cells that are responsible for adaptive and innate immune responses. The myeloid–erythroid CMP pathway gives rise to erythroids, granulocytes, monocytes and megakaryocytes.

They reported information on the key cell surface marker transitions during the passage of transformation of HSC to the generation of MPPs. Human HSCs were shown to be CD45RA−, CD90+, CD49f+; MPPs were CD45RA−, CD90− and CD49f−. Increased expression of transcription factor genes MYC, IKZF1 was found to be associated with the MPP state.

Doulatov *et al.* (2012) documented additional changes in cell surface antigens as cells differentiated to MLPs. The MLPs were CD45RA+, CD10+ and CD7−. As cells transitioned from MPPs to myeloid–erythroid progenitors (CMP cells), antigens changes so that cells were CD45−, CD135+, CD10− and CD7−.

The MLPs give rise through defined replication stages to T, B and NK cells. The CMP cells give rise to granulocytes, monocytes and megakaryocytes. Doulatov *et al.* (2012) indicated that dendritic cells might be derived from MLPs and/or CMP cells.

Key transcription factors involved in the differentiation of MPPs to the MLP series and T and B cells include GATA2 and PU1 (also known as SPI1 ETS transcription factor), Transcription factors involved in the transition from MPP cells to the CMP series included PU1 and BCLIIA.

Laurenti *et al.* (2013) analyzed gene expression profiles of hematopoietic cell subtypes isolated from cord blood. They reported that early progenitor cells have the capacity to give rise downstream to several different differentiated cells. Transcription factors are essential to differentiation processes. Transcription factors interact with DNA to modify expression of target genes and chromatin structure and modifications of DNA impact transcription factor binding. GATA1 and GATA2 transcription factors are important in further differentiation of cells derived from megakaryocyte erythroid progenitors.

Laurenti *et al.* (2013) reported that transcription factors important in differentiation of B-lymphocytes from the common lymphoid progenitor included BCL11A, SOX14 and TEAD1 (transcriptional enhancer factor). BCL6 is important in differentiation of several lymphoid cell types.

Therapeutic Transplants of Bone-Marrow

For the past 50 years extensive clinical experience in therapeutic application of bone-marrow transplantation has been gained. However bone narrow samples that are immunologically matched to a specific patient are in short supply. Therefore increases in availability of *in vitro* cultured cells are important.

Studies on maintenance of *in vitro* cultures of stem cells are important and studies on aspects of stem cell regeneration and differentiation of these cells are critical. Gathering of this knowledge depends on analysis of both embryonic stem cells (ESCs) and induced pluripotent stem cells (iPSCs) (Cherry and Daley, 2013). These authors noted that for transplantation, it is essential that differentiated cells be administered since in the presence of undifferentiated cells, iPSC may increase the likelihood of cancer. Furthermore, when iPSCs are used, these cells cannot be derived through the use of retroviruses containing the induction factors.

The advantage of using iPSC derivation procedures is that stem cells can be derived from the patient. Furthermore in the case of genetic diseases correction of the gene defect can be carried out in the iPSCs.

Insight into factors that determine differentiation has been obtained through *in vitro* culture studies of hematopoietic precursors and on human pluripotent stem cells.

There are separate aspects, the first is to derive multi-potential hematopoietic cells and the second is derivation of differentiated cells; a third step involves engraftment of cells. A number of investigators have developed efficient *in vitro* culture procedures to propagate HSCs and to derive differentiated cells (Woods *et al.*, 2011; Kennedy *et al.*, 2012). Culture conditions to achieve directed differentiation of HSCs include addition of morphogens including activin A, BMP4 Notch ligand. However, the number of transplantable cells derived from these cultures were usually found to be low.

CHAPTER 11

ABDOMINAL WALL AND GASTRO-INTESTINAL TRACT

Abdominal Wall Development

By the fourth week of human embryonic life folding of the embryonic disc occurs so that the lateral edges meet in the midline. The outer ectoderm covers the entire surface of this tube except for the umbilicus and the yolk sac (Larsen, 2009). This folding process is followed by development of the gut tube. Larsen (2009) described this as a tube within a tube. There is subsequently a secondary stage of abdominal wall development that includes connective tissue and myoblast infiltration in the ectoderm. Myoblasts differentiate to produce fibers and there is a specific orientation and patterning of the muscle fibers.

Nichol *et al.* (2012) reviewed this secondary abdominal wall formation in normal fetuses and in cases with abdominal wall defects. Their studies revealed that in omphalocoele, there was lack of unidirectional alignment and reduced migration of myoblasts to the ventral midline. In addition there was lack of unidirectional alignment of fibers in specific muscles. Thinning of connective tissue between the muscle layers was also observed.

Studies in the mouse revealed the factors important in development of the secondary abdominal wall in mice, which include Msx1, Msx2 (muscle segment homeobox transcription factors), IGFII (IGF2 insulin like growth factor), and FGFR1 and 2 (fibroblast growth factor receptors). Mutations in ROCK1 (Rho protein kinase) that impacts actin organization have also been implicated in causation of omphalocoele in mice.

Anterior Abdominal Wall and Defects

Gastroschisis and omphalocoele result when the anterior abdominal wall fails to develop appropriately. In omphalocoele, the gut is covered with membrane and in gastroschisis, it is not covered by membrane.

The etiological factors that predispose to abdominal wall defects are not known at the time of writing. However, there is evidence for a rising incidence of this birth defect in the United States and in the United Kingdom and in sub-Saharan Africa. Keys *et al.* (2008) reported that in the United Kingdom, there was 3-fold increase in incidence over the 10 years prior to 2008. Baerg *et al.* (2003) reported that in Canada the incidence of gastroschisis rose from 1.35 per thousand in the period 1985–1990 to 4.06 per thousand in the period 1996–2000.

The greatest known risk factor includes young maternal age and first pregnancy. Lubinsky (2012) reported that pregnancy in very young women is associated with increased estrogen levels. High estrogen levels predispose to thrombophilia and to thrombosis. Lubinsky (2012) also drew attention to the increased environmental estrogen levels that predispose to thrombophilia and thrombosis. He noted that the increase in environmental estrogen parallels the increase in incidence of gastroschisis.

Lubinsky (2014) proposed a vascular thrombotic pathogenesis for gastroschisis. He proposed specifically that involution of the right umbilical vein created thrombosis adjacent to the umbilical ring.

Zhang *et al.* (2014) reported that ectodermal WNT signaling induces development of abdominal musculature from mesoderm. They reported that loss of WNT results in dysgenesis of anterior (ventral) abdominal musculature.

Gastro-Intestinal Tract

In early embryogenesis, the gastro-intestinal tract exists as a blind-ended tube and the mid-gut opens into the yolk sac. Vascularization of the abdominal gut occurs from the celiac arteries and the mesenteric arteries. The abdominal portion of the esophagus, the stomach and the upper (proximal) duodenum are detectable by the fifth week (Larsen, 2009).

The midgut differentiates to distal duodenum, jejunum, ileum and ascending colon and two-thirds of the transverse colon. The hindgut gives rise to the distal transverse colon, the descending and sigmoid colon and rectum. The cloaca at the distal end of the gut tube forms the anorectal canal and the urogenital sinus.

Anorectal malformations

These malformations occur with frequency of 1 per 2500 births. The caudal hindgut forms the anorectal tract and part of the genito–urinary system. Guo *et al.* (2014) reported that asymmetric growth of the cloacal mesenchyme drives the re-shaping of the cloaca and separates the anorectal and genito–urinary tracts. In studies in mice they identified a signaling molecule Dkk1 that is highly expressed in the peri-cloacal mesenchymal progenitor cells. Deletion of Dkk1 predisposed to imperforate anus and recto–urinary fistula in mice. They determined that Dkk1 deletion led to increase in WNT activity in the peri-cloacal mesenchyme. Dkk1 signaling is important in inhibiting WNT signaling at an appropriate time in development.

Enteric nervous system

Enervation of the bowel is provided by the enteric nervous system (ENS). Lake and Heuckeroth (2013) reviewed the role of the neural crest in ENS development. Steps in the process include neural crest cell invasion of the intestinal wall. Guidance factors and morphogens then determine migration and differentiation.

Impaired development of the ENS leads to aganglionosis, Hirschsprung disease and intestinal obstruction. It is important to note that in Hirschsprung disease, the extent of anganglionosis varies between individuals. Hirschsprung disease involves the rectum and the sigmoid colon and the length of the segment, i.e., aganglionic, differs in different individuals.

Hirschsprung disease most frequently occurs sporadically and a large number of different factors and gene activities contribute to appropriate enervation and function of the musculature of the colon and rectum.

Butler-Tjaden and Trainor (2013) reported that key genes that regulate neural crest cell development and migration which are involved in Hirschsprung disease include RET (receptor tyrosine kinase), GDNF (glial cell derived neurotrophic factor), GFRA1 (GDNF receptor), NRTN (neuturin), EDNRB (endothelin receptor type B), ET3 (EDN3 endothelin3), ZFHX1B (zinc finger homeobox transcription repressor), PHOX2B (homeodomain transcription factor), SOX10 (transcription factor) and SHH (tissue patterning and cilia function). They also noted however, that these genes and proteins were not mutated in at least 50% of cases and they proposed that other proteins and modifiers play roles.

ET3 or EDNRB defects are in some cases associated with Waardenburg syndrome type 4. In this syndrome, pigmentation defects, sensorineural deafness and aganglionic megacolon may occur. This syndrome may also arise due to *SOX10* mutations (Bondurand *et al.*, 2012).

PHOX2B is a homeodomain transcription factor expressed in neural crest cells and in the autonomic nervous system. PHOX2B mutations in humans, particularly polyalanine repeat expansions lead to congenital central hypoventilation syndrome (Ondine's curse, where patients fail to breathe adequately during sleep. Lake and Heuckeroth (2013) reported that PHOX2 mutations might also lead to the Hirschsprung disease.

Other transcription factors involved in neural crest cell migration and ENS development include paired box transcription factor PAX3 and ZEB2 (a zinc finger transcriptional repressor), and mutations in these products may also lead to aganglionosis.

The hedgehog signaling proteins, SHH and IHH, act as morphogens for development of the ENS. Hedgehog signaling induces bone morphogenic protein BMP4 which is important for ENS patterning. Retinoic acid is also an important morphogen for ENS development.

Congenital diaphragmatic hernias and tracheal–esophageal fistulas are discussed in the chapter on lung and diaphragm development.

CHAPTER 12

LUNG AND DIAPHRAGM DEVELOPMENT

Stages in Lung Development

Lung development in humans begins on day 22 post-conception with out-growth of a bud from the ventral surface of the endoderm of the foregut and an out-pouching of the surrounding mesoderm. In humans, lung development is only completed during adolescence (Larsen, 2009).

Early stages include further outgrowth and elongation of the respiratory bud and initial branching to form the right and left primary bronchi. The proximal end of the respiratory bud gives rise to the trachea and larynx. Impaired proliferation of cells of the proximal end of the respiratory bud or impaired separation from the underlying gut endoderm that gives rise to the esophagus may lead to the development of trachea–esophageal fistulas. There is evidence from studies in mouse that the transcription factor NKX2.1 (also known as thyroid transcription factor) and the sonic hedgehog signaling pathway play key roles in determining appropriate development of the lung bud, its proximal end and separation from the esophagus forming region.

The stage of lung morphogenesis during which the primary bronchial branches are formed and undergo further branching to give rise to a net-work of branches is referred to as the pseudo-glandular stage. In the next stage, the canalicular stage, further branching of the network occurs and this is associated with angiogenesis and vascularization. The following stage of development is referred to as the saccular stage; during this stage differentiation of the alveolar epithelium occurs along with surfactant synthesis and development of the vascular capillary network. Further sub-divisions of saccules and full maturation of the alveoli occur during the alveolarization stage (Hagood and Ambalavanan, 2013).

Signaling pathways and transcription factors in respiratory component development

Extensive studies on signaling pathways and transcription factors involved in specification and lineage determination of respiratory components have been carried out particularly in mice but also in humans (Herriges and Morrisey, 2014). Studies have revealed that expression of the transcription factor NKX2.1 is the first evidence of respiratory bud specification in a specific region of the ventral anterior foregut endoderm. Multiple different isoforms are derived from transcripts of the NKX2.1 gene.

Expression of NKX2.1 occurs in response to WNT2/beta-catenin signaling from the mesenchyme that overlies the specific segment of the gut ventral endoderm. Branching morphogenesis is particularly dependent on fibroblast growth factor FGF10. However interactions between FGF10 and SHH are also likely important. Development of the epithelial sheet within bronchial branches is likely impacted by components of the planar cell polarity pathway and FGF and WNT signaling components impact orientation of the cells.

During subsequent stages, interactions of the epithelial cells with the matrix are important. These interactions are in part dependent on fibronectin and its interaction with integrin receptors on cells. With further development proximal and distal epithelial cell lineages take on different characteristics. Transcription factors that play roles in epithelial differentiation include GATA6, NKX2.1, FOXA1 and FOXA2. Herriges and Morrisey (2014) reported that the proximal progenitor cells under influence of transcription factor SOX2 give rise to neuro-endocrine, secretory ciliated and mucosal cells. One class of secretory cells that has been extensively analyzed are the Clara cells that produce secretoglobin. Differentiation of these cells is dependent on NOTCH signaling. NOTCH proteins are present on cell surfaces and act as receptors for multiple ligands.

The distal progenitors differentiate into type 1 and type 2 alveolar cells, AEC1 and AEC2, under influence of ID2 (inhibitor of DNA binding), a transcription regulator and transcription factors NMYC and FOXA1 and FOXA2. Subsequent lung alveolarization requires ETS transcription factor.

In addition to the development of the different epithelial cell lineages mesodermal derivatives are generated and these include airway and vascular smooth muscle and pericytes that wrap around venules and capillaries. FGF10 likely plays important roles in mesenchyme proliferation and generation.

There is also evidence that cardiopulmonary progenitor cells contribute to the generation of mesodermal lineages in the lung including vascular and airway smooth muscle. These cells express WNT2, GLI (zinc finger transcription factors) and ISL1 (insulin gene binding transcription factor).

Epigenetic factors including histone modification likely play important roles in regulation and differentiation.

Surfactant

The type 1 and type 2 alveolar cells (AEC1 and AEC2 cells) synthesize and secrete surfactant. Surfactant is a mixture of 90% lipids and 10% protein. Hydrophilic proteins in surfactant are protein A and D. Surfactant A protein is encoded by two genes SFTPA1 and SFTPA2 and surfactant protein D is encoded by one gene, SFTPD. In addition surfactant contains two hydrophobic proteins surfactant B and surfactant C (Gower and Nogee, 2011). Surfactant is packaged into lysosomal derived organelles, lamellar bodies. The transport protein, adenosine triphosphate binding cassette protein encoded by the ABCA3 gene is essential for moving surfactant phospholipid to lamellar bodies. Surfactant is then secreted and it adsorbs onto the inter-space that separates alveolar air from liquid. Surfactant lowers the surface tension. Genes that encode surfactant proteins respond to transcription factor NKX2.1 and expression occurs later in gestation.

Deficiency of surfactant proteins in premature infants predisposes them to respiratory distress syndrome. Mutations in the gene that encodes surfactant B and impair production of that protein lead to respiratory distress syndrome in full-term infants. Mutations in the surfactant C encoding gene (SFTPC) and in the ABCA3 transporter encoding gene may also lead to impaired respiratory function. However, this impaired function often presents later in childhood or even in young adults.

Stem Cells and Airway Progenitors

Generation of airway progenitors from mouse embryonic stem cells and from human induced pluripotent stem cells (iPSCs) was achieved in 2012

(Mou *et al.*, 2012). These investigators converted iPSCs into lung endoderm using transcription factor NKX2.1 and precisely timed signaling with activin, BMP and WNT.

Diaphragm and Defects in Development

The diaphragm is a musculo–tendinous structure derived from four different structures: the septum transversum, the pleuroperitoneal membranes, the body wall mesenchyme and the esophageal mesoderm. It is innervated by the phrenic nerve (Larsen, 2009). Diaphragmatic hernias can originate at different position and can vary in size. The most common defect is a left sided defect that fails to seal off the pleural cavity from the peritoneal cavity. Smaller defects also occur and these are frequently in the parasternal region. With large defects abdominal contents herniate into the pleural cavity and compromise cardiac and lung function.

In humans, this birth defect occurs with a frequency between 1 in 3000 and 1 in 4000. Congenital diaphragmatic hernia (CDH) is also often associated with defects in the pulmonary airway and defects in vascular development (Ackerman and Pober, 2007). Postero-lateral defects are more common and anterior defects are less frequent.

Deficiency of Vitamin A (retinol) and defects in retinoic acid receptors and retinoic acid signaling were reported in cases of CDH. Mutations in retinoid signaling related genes and/or transcription factors occurred in some cases of the disorders. These genes include transcription factors FOG2 and GATA4. Other gene products that are implicated include retinol binding or in retinol metabolism STRA6 (membrane protein involved in retinol metabolism), LRAT (acetyl transferase for esterification of retinol), CRBP1, CRBP2, RBP1 and RBP2 (retinol binding proteins).

Beurskens *et al.* (2010) analyzed levels of retinol and RBP in maternal blood and newborn cord blood samples of 22 cases with CDH, and 34 controls. They determined that newborns with CDH had significantly lower levels of retinol than control newborns. However, retinol and RBP levels were not lower in mothers of cases than in mothers of controls. They concluded that CDH was due to altered homeostasis in the fetus/ newborn cases independent of maternal status.

Goumy *et al.* (2010) reported that 10–20% of cases of CDH have chromosome abnormalities. Microdeletions of chromosome 8p23.1 have been described in several cases with CDH and this chromosomal region harbors the GATA4 and NEIL2 genes that likely play role in CDH (Keitges *et al.*, 2013). The NEIL2 gene encodes a glycosylase. There is also evidence that specific mutations in the GATA4 transcription factor can lead to CDH.

Slavotinek (2014) reported that the etiology of CDH is not defined in more than 50% of the patients with this abnormality. The most frequent chromosome abnormalities found in this condition include deletions of 15q26, 8p23, 4p16.3, and tetrasomy 12p. CDH also occurs in a sub-set of patients with translocation *t* (11; 22) (q23:q11).

CDH may also occur as one of the abnormalities present in more than 70 different malformation syndromes. They are also more common in fetuses with overgrowth syndromes including Beckwith Wiedemann syndrome and Simpson–Golabi–Behmel syndrome. Slavotinek (2014) noted that there is evidence that vitamin A deficiency predisposes to CDH.

CDH sometimes occurs in association with other birth defects, e.g., in Fryns syndrome where CDH occur along with oral–facial clefts, skeletal anomalies and organ abnormalities. A number of cases with this syndrome have been found to have microdeletions on chromosome 1q41–1q42.

CHAPTER 13

LIVER AND PANCREAS DEVELOPMENT

Liver Development

In humans, liver development begins on approximately day 22 of fetal life when an endodermis thickening forms on the ventral side of the embryonic foregut (Larsen, 2009). The pancreas arises from the dorsal side of the embryonic foregut. Insights into the steps of development have been obtained through studies on vertebrates especially mouse.

Si-Tayeb *et al*. (2010) reviewed organogenesis and development of the liver. Inductive signals for differentiation of hepatic progenitor cells include fibroblast growth factors (FGF1 and FGF2). These signals then trigger activity in mitogen-activated protein kinase (MAPK) signaling pathways. There is also evidence that transcription factor GATA and bone morphogenic protein BMP4 are important in induction of the liver bud. They are expressed from cells in the septum transversum that separates the thoracic and abdominal cavities. Hepatic specification of the foregut endoderm is also promoted by the homeobox transcription factor HHEX.

Differentiation of hepatic progenitor cells (HPCs) that express albumin is promoted by the zinc finger transcription factors FOXA1 and FOXA2 and transcription activators GATA4, hepatic nuclear factor 1B (HNF1B) and HNF6 (also known as Onecut, OC1). Metalloproteinase (MMP) facilitates migration of HPCs into the liver bud stroma. Differentiation of progenitor cells to hepatoblasts and their migration requires a network of transcription factors. Transcription factor PROX1 (Prospero homeobox) promotes hepatoblast proliferation and migration.

Subsequent differentiation of hepatoblasts to hepatocytes requires hepatic nuclear factors HNF1alpha, HNF1beta, HNF4alpha, HNF3, NR5A2 and transcription factor FOXAA2. WNT beta-catenin signaling is necessary for positioning of hepatocytes within the liver lobule. Fetal hepatocytes express several genes that are not expressed in hepatocytes later in life; these include alpha-fetoprotein (AFP) and GYP3 (glypican 3, a membrane proteoglycan).

Lee *et al.* (2012) analyzed ontogeny in human and rodent liver. They reported that hepatoblasts that expressed AFP, albumin and transthyretin (carrier protein) are bi-functional. When located next to portal veins these hepatoblasts give rise to epithelial cells that line the intra-hepatic bile ducts. Hepatoblasts located within the liver parenchyma differentiate to hepatocytes.

SOX9 homeobox transcription factor and transforming growth factor beta (TGFbeta) expression are necessary for differentiation of hepato-blasts to cholangiocytes (epithelial cells of the bile ducts) (Si-Tayeb *et al.*, 2010). There is evidence that NOTCH receptor signaling and expression of the NOTCH ligand JAG1 (Jagged1) play important roles in biliary development and was obtained through investigation of the human mal-formation syndrome, Alagille syndrome. One feature of this syndrome is that few intra-hepatic bike ducts are present. There is also evidence that cholangiocyte proliferation and growth of bile ducts requires function of cilia. Defects in intra-hepatic ducts are sometimes seen in cases with defect in the conditions PKHD1 (polycystic kidney disease and hepatic disease) and PKD2 (autosomal dominant polycystic kidney disease) (see section on Cilia).

Other important aspects of liver development include differentiation of stellate cell that store vitamin A, and differentiation of endothelial cells that line sinusoids and stroma proliferation.

Fetal Liver and Hematopoiesis

Erythroid precursors generated from the yolk sac enter the liver. These erythroid precursors give rise to islands of proliferation cells and the liver is the major site of hematopoiesis, prior to the shift to the bone-marrow (Lee *et al.*, 2012).

Analyses of Stem Cell Differentiation into Liver Cells

Stages in the development of hepatocytes from pluripotent stem cells (PSCs) mimic the stages of embryonic development of the liver (Hannan *et al.*, 2013). The first stage involves differentiation of induced pluripotent stem cells (iPSCs) to definitive endodermal cells. In the second stage definitive endodermal cells differentiate to anterior definite endoderm cells (ADE). In a third stage the ADE cells differentiate to liver cell progenitors. The fourth stage involves functional maturation of the liver precursor cells, hepatoblasts, to generate hepatocytes, liver-like cells.

Hannan *et al.* (2013) used defined culture media and specific growth factors to achieve these developmental stages. In addition they monitored gene expression signatures of the cells at different stages. Endoderm specification and commitment requires activin A, FGF and BMP4. Liver cell commitment requires BMP4 and FGF10. Hepatocyte maturation required hepatocyte basal medium (HBM), hepatocyte growth factor (HGF) and oncostatin M (OSM). HGF and OSM are cytokine like proteins.

Monitoring of stage specific markers revealed that definitive endoderm cells expressed SOX17 and CXCR4 (chemokine receptor). Anterior definitive endoderm cells expressed FOXA2 and HNF1B. Specification of hepatoblasts was characterized by expression of HNF4A, PROX1, HHEX, AFP and TBX3.

Maturation of hepatoblasts into hepatocyte like cells was characterized by expression of AFP and albumin. Subsequently, cells manifest cytochrome p450 activity.

Human hepatocyte transplantation is increasingly being utilized to treat liver failure (Fitzpatrick *et al.*, 2009).

Analysis of Developmental Phase Specific Gene Expression

Smith *et al.* (1971, 1972, 1973a, 1973b) described developmental stage specific and tissue changes in expression of alcohol dehydrogenase enzymes. The ADH1 (ADH1A based on new nomenclature) gene that encodes the alpha isozyme is the only ADH form expressed in fetal liver during the first trimester. During the second trimester the ADH1A and ADH1B genes are expressed. In post natal life all three genes ADH1A, ADH1B and ADH1C are expressed. The ADH1B gene is expressed in a number of different tissues. The ADH1C gene is primarily expressed in kidney and in the gastro-intestinal tract. The three genes ADH1A, ADH1B and ADH1C are currently defined as the class I ADH genes.

Detailed analyses have been carried out to define factors that determine class I ADH gene regulation (Dannenberg *et al*., 2005). They reported that the transcription factor GATA2 played an important role in controlling this expression.

Van Ooij *et al*. (1992) reported that the hepatic nuclear factor1 (HNF1) controlled ADH1A gene expression and the ADH1A promoter responded to HNF1. Su *et al*. (2006) identified a DNAse1 hypersensitivity site 51Kb upstream of the class 1 ADH genes that bound HNF1 and played a key role in transcriptional activation of expression of these genes in liver. This 275bp fragment displayed DNAse1 hypersensitivity that was liver specific. In transgenic mice, into which the human ADH gene region was cloned, deletion of the 275bp conserved region led to shutdown of expression of the class 1 ADH genes in liver.

Angiocrine Signals in Liver Regeneration

There is evidence that vascular endothelium and angiocrine signals play important roles in development, in tissue homeostasis and in tissue regeneration. Hu *et al*. (2014) reported that liver vascular sinusoidal endothelial cells play important role in promoting hepatocyte proliferation during liver regeneration. They demonstrated that angiopoietin 2 (ANGPT2) is dynamically regulated after partial hepatectomy.

After hepatectomy, there is an inductive phase during which hepatocyte numbers increase. In that phase, angiocrine production is inhibited. Cyclin D1 expression promotes hepatocyte proliferation.

In the subsequent phase, vascular endothelial growth factor receptor 2 (VEGFR2) expression increases. This leads to regenerative angiogenesis. Hu *et al*. (2014) reported that the liver sinusoid endothelial cells orchestrate regeneration.

Hyperbilirubinemia, Cholestasis, and Liver Dysfunction in Newborns and Young Infants

Bilirubin

Unconjugated bilirubin is derived from hemoglobin released during the breakdown of red blood cells in the reticulo–endothelial system.

Unconjugated bilirubin then circulates in the blood, bound in part to albumin. It passes into hepatocytes. The solute carriers OATP1B1 (SLCO1B1) and OATP1B3 (SLCO1B3) are members of the SLCO family of organic solute carriers which facilitate uptake of bilirubin by hepatocytes (Erlinger *et al.*, 2014). A specific protein (ligandin) in hepatocytes has affinity for bilirubin. Within hepatocytes conjugation of bilirubin with glucuronic acid occurs. The conjugated bilirubin then passes from hepatocytes to bile canaliculi.

Hyperbilirubinemia with unconjugated bilirubin in newborns is often related to increased hemolysis. However, increased levels of bilirubin in blood can result from defective uptake of unconjugated bilirubin by hepatocytes, defective conjugation of bilirubin to glucuronide, or defective secretion of bilirubin into canaliculi.

In Gilbert syndrome, there is mild hyperbilirubinemia involving unconjugated bilirubin, without evidence of hemolysis. Gilbert syndrome results from mutation in the promoter region of *UGT1A* gene that encodes Uridine diphosphate glucuronyl transferase. Homozygotes for these mutations have enzyme levels approximately 10% of those of controls.

Crigler Najjar syndrome is a rare recessive disorder with high levels of unconjugated bilirubin. This disorder presents itself in newborns, potentially leading to kernicterus, deposition of unconjugated bilirubin and damage to brain nucleus. This condition arises due to mutations in exons 2 and 5 of *UGT1A*.

Dubin–Johnson syndrome is characterized by hyperbilirubinemia and increased precipitation of a melanin like pigment in liver. This condition is due to mutations in the ATP dependent transporter protein ABCC2 that is expressed in hepatocyte membranes and elsewhere. In Dubin–Johnson syndrome, there are defects in the capacity of the hepatocytes to secrete conjugated bilirubin into the bile and into the canaliculi.

Bile components and bile salt synthesis

Bile is composed of 98% water, bile salts, bilirubin, cholesterol and lipids. Bile is pumped into the canalicular structures in the liver. Bile salts act as detergents that facilitate the movement of bile.

Bile acids are derived through enzymatic transformation of cholesterol in a series of reactions that utilize cytochrome p450 enzymes. Clayton (2011) reviewed disorders of bile acid synthesis and noted that these disorders can lead to life threatening cholestatic liver disease in infancy but can also manifest symptoms later in life.

High blood levels of conjugated bilirubin and raised levels of liver transaminases characterize cholestatic disease in infancy. However the levels of gamma glutamyl transpeptidase are normal. Clayton (2011) noted that bile acids and bile salts facilitate release of gamma glutamyl transpeptidase from the liver canalicular membranes. In the absence of bile acids and bile salts this release does not occur. In this form of cholestatic liver disease in infancy the liver histological features are defined as Giant Cell Hepatitis.

Conversion of cholesterol to bile acids

This requires modification of the cholesterol side chain and of the cholesterol nucleus. Bile acid synthesis occurs along two major pathways, the neutral pathway that is the most commonly used pathway in adult life and the acidic pathway that is the most important pathway during the first year of life (Clayton, 2011).

In the neutral pathway, cholesterol is first converted to 7-hydroxy cholesterol through action of the enzyme CYP7A1 (cholesterol-7-alpha hydroxylase). Progressive conversion including hydroxylations and oxidations in peroxisomes, leads to production of cholic acid and chenodeoxycholic acid.

The acidic pathway starts with conversion of cholesterol to 27-hydroxycholesterol through the activity of CYP27A1 (sterol-27-hydroxylase). Subsequent reactions lead to the production of chenodeoxycholic acid. Clayton (2011) reported that the neutral pathway of bile acid synthesis occurs only in liver. The acidic pathway occurs in liver and can occur in other tissues.

Synthesis of bile acid is suppressed by high levels of bile acids in the enterohepatic circulation and is stimulated by high levels of cholesterol. Cholesterol stimulation acts *via* the transcription factors, liver X receptor (LXR) and hepatocyte nuclear factor to induce synthesis of cholesterol-7-alpha-hydroxylase). In the acid pathway, cholesterol stimulates bile acid synthesis *via* the 27-hydroxy cholesterol pathway.

Neonatal hyperbilirubinemia and cholestasis also occurs in patients with deficiencies of the enzyme HSD3B2 (hydroxy-delta-5-steroid dehydrogenase, or HSD3B7 3 beta- and steroid delta-isomerase 7).

Bile acid conjugation with amino acids glycine and taurine is catalyzed with the product of the BAAT gene locus (bile acid amino acid N acetyl transferase). Deficiency of this enzyme occurs in the Amish populations and leads to familial hypercholanemia.

Liver Failure in Mitochondrial Depletion Syndrome

Al-Hussaini *et al.* (2014) described mitochondrial depletion, leading to cholestasis and liver failure in newborns and infants, three months of age or younger. In their study in Saudi Arabia of 450 patients with infantile cholestasis, 20 of the patients had manifestations of hepatocerebral mitochondrial depletions syndrome. In 11 of the patients, pathogenic mutations occurred either in DGUOK (dioxyguanosine kinase) or in MPV17 (a mitochondrial inner membrane protein). These patients manifested hepatomegaly, mild splenomegaly and in addition had hypotonia; two patients had seizures. The authors noted that in all of these patients, the parents were consanguineous. Mutations in MPV17 have been reported in patients from other ethnic groups, including patients with Navajo neurohepatopathy.

Pancreas Development

The pancreas develops from dorsal and ventral buds of the foregut endoderm. Benitez *et al.* (2012) reviewed pancreatic development. Pancreatic budding is dependent upon expression of transcription factors PTF1A (pancreas specific transcription factor), PDX1 (pancreatic and duodenal homeobox), and MYC (phosphoprotein transcription factor). The differentiation begins with proliferation of pancreatic progenitors of the foregut endoderm. A lumen develops within the proliferating cell mass and this subsequently gives rise to a tubular structure with a stalk domain and a tip. Benitez *et al.* (2012) reported that cells in the stalk are bipotent, while cells located in the tip are multipotent. These multipotent cells subsequently give rise to acinar cells and ductal cells in the pancreatic exocrine system and to cells of the endocrine system. The bipotential cells of the stalk give rise to endocrine cells and pancreatic duct cells.

Differentiation of exocrine pancreatic elements

Continued PTF1A expression is important for differentiation of the acinar precursor cells from the multipotent progenitor cells. Benitez *et al.* (2012) emphasized that acinar cell generation is also dependent upon factors secreted from mesenchymal cells that surround the bud. These cells produce FGF10 growth factor and they produce Follistatin that inhibits TGFbeta signaling, which if expressed inhibits differentiation.

Maturation of the acinar cells that are responsible for secreting hydrolytic enzymes into the duodenum involves proliferation of mitochondria, proliferation of the Golgi and endoplasmic reticulum systems and development of exocrine granules. This maturation is dependent upon the MIST1 (BHLHA15 basic helix loop helix) transcription factor. Maturation is also dependent upon the formation of a trimeric complex that contains PTF1, TCF12 (transcription factor 12) and RBPJ (recombination related transcriptional regulator). This complex is referred to as PTF1-J.

Exocrine duct cells secrete mucins and bicarbonate. Benitez *et al.* (2012) noted that definitive regulators of duct cell development were not characterized. However, there is evidence that development of cilia is essential for duct function and cilia development requires expression of nuclear receptor transcription and HNF1B and HNF6 (OC1).

Differentiation of Endocrine Pancreatic Elements

The endocrine progenitor cells are derived primarily from the stalk domain of the pancreatic bud. Production of NGN3 (neurogenin 3) is critical for development of these cells. A number of transcription factors downstream of NGN3 are essential for differentiation of the different types of pancreatic endocrine cells. These cells include the alpha glucagon producing cells, beta cells that produce insulin, PP cells that produce pancreatic polypeptide, the delta cells that produce somatostatin and the epsilon cells that temporarily produce ghrelin.

Differentiation of the alpha cells requires transcription factors PAX6 (paired box), MAFB (leucine zipper), ARX (homeobox) and FOXA2 (Forkhead box nuclear factor). The PAX4 (paired box transcription factor) is required for differentiation of the somatostatin producing delta cells.

Differentiation of beta cells that produce insulin requires a number of different transcription factors including MAFB, PDX1 (pancreatic and duodenal homeobox), PAX4, PAX6 and NKX6-1 homeobox transcription factor. Benitez *et al.* (2012) reported that for beta cell development, it is important that ARX be silenced and there is evidence that silencing of expression of this gene is dependent upon activity of methyltransferase DNMT1 and DNMT2.

Meier *et al.* (2010) studied human embryos and fetuses between eight week post-conception and birth. They determined that beta cells were present in the pancreas from week 9 onward and increased in mass in a linear fashion until birth. Endocrine cells were initially adjacent to primitive ductal epithelial cells. After nine weeks, beta cells were present in islets. They reported that insulin and glucagon were produced in the early fetal period. In the early post-natal period, high rates of beta cell replication occurred.

Post-natal maturation of beta cells

Benitez *et al.* (2012) reported that post-natal beta cell maturation is characterized by changes in expression of a number of enzymes which function in metabolism. Lactose dehydrogenase isozyme A (LDHA) shows high expression in immature B cells and there is evidence that glycolysis rather than oxidative metabolism predominates prior to maturation, and production of GLUT2 (SLC2A2 glucose transporter), glucokinase and PCSK1/PCSK3 proconvertase is also lower. Maturation of B cells is dependent on expression of PDX1 and NEUROD1 transcription factors. PDX1 haploinsufficiency leads to a form of diabetes mellitus. Early in the post-natal period, beta cells undergo proliferation.

Pancreatic beta cells contain many mitochondria and in addition dense core granules. Pro-insulin is initially contained in immature secretory granules. The maturation process in the granules involves processing of pro-insulin by proconvertases and carboxypeptidases.

Ghrelin

Ghrelin is expressed in numerous cell types, and particularly in the stomach. However, it is also expressed in an independent cell type in the pancreatic islands. Kawamata *et al.* (2014) determined that concentrations of leptin (produced by fat cells), ghrelin, amylin (islet amyloid polypeptide), insulin polypeptide, and glucagon like peptide changed in the postnatal period.

Pancreas Regenerative Therapies for Diabetes

Rodriguez-Segui *et al.* (2012) emphasized that understanding the mechanisms of development and differentiation of the pancreas are central to development of regenerative therapies for diabetes. They reported that insight into these processes have come from studies in mice. Key transcription factors in mice include Pdx1, Pt1a, Mnx1 (motor neuron and pancreas homeobox), Sox9 and Hnf1b. Gene mutation or gene knockout studies revealed the importance of Foxa1/a2, Onecut1 (Hnf6) and Nkx6.1. They noted that mutations in GATA6 and GATA4 transcription factors have been shown to lead pancreas agenesis in humans. Rodriguez-Segui *et al.* (2012) noted however, that specific mutations in patients with monogenic form of diabetes leading to impaired pancreatic development provided evidence for the important roles of PDX1, PTF1A and HNF1B.

Based on studies in mouse and human Rodriguez-Segui *et al.* (2012) proposes that the toolkit for pancreas genesis includes: GATA6, GATA4, PDX1, FOXA1, FOXA2, ONECUT1, HNF1B, MNX1, SOX9, NKX6.1 and NKX6.2. They reported that inactivation of each one of these factors separately or in combination impact pancreatic progenitor proliferation and or differentiation.

Khoo *et al.* (2012) reported that Pdx1 is a master regulator of pancreas development. Pdx1 (pancreas duodenal homeobox, also known as insulin promoter factor 1) interacts with Neurod1 transcription factor to regulate insulin production in beta cells. Studies on Pdx$^{+/-}$ mice have impaired insulin secretion and increased beta cell apoptosis. Pdx1 activates transcription of a number of different genes including the genes that encode insulin, somatostatin, glucokinase, islet amyloid polypeptide and glucose transporter.

Gauthier *et al.* (2004) reported that Pdx1 is an essential regulator of mitochondrial metabolism and that Pdx1 deficiency causes mitochondrial dysfunction.

Induction of Insulin Production in Embryonic Stem Cells (ESCs)

Wei *et al.* (2013) induced human ESCs into insulin producing cells in a step-wise procedure which resembled the one involved in differentiation of pancreatic cells from definitive gut endoderm. They determined that regulated expression of specific microRNAs was important in this process. Key factors involved in derivation of pancreatic progenitor cells from definitive endodermal cells including Sox 17, FoxA2 and miR34a. Differentiation of pancreatic progenitor cells to insulin producing cells required factors Hnf1b, Ngn (neogenin), Sox9, Pdx1, Isl1 (insulin enhancer lim homeobox), miR146a, miR7 and miR374.

In addition to growth factor and transcription factor expression, epigenetic factors are also important in determining differentiation. Avrahami and Kaestner (2014) carried out phenotypic studies in mice with mutations in histone modifying enzymes and in the presence of activators or inhibitors of acetylation and methylation. They emphasized that understanding the epigenetic status of pancreatic beta cell is important to determine epigenetic conditions under which insulin is produced. iPSC and ESC derived cells frequently do not produce insulin based on glucose levels.

CHAPTER 14

BONE AND EXTRA-CELLULAR MATRIX

Intra-Membranous and Endochondral Ossification

Distinct processes in bone formation include intra-membranous ossification and endochondral ossification. Mackie *et al.* (2011) reviewed these processes. They noted that the flat bones of the skull are derived by intra-membranous ossification and in this process embryonic mesenchymal cells give rise directly to osteoblasts that form bone.

Endochondral bone formation is involved in the formation of most of the skeleton and in this process embryonic mesenchymal cells differentiate to chondrocytes. These in turn secrete extra-cellular matrix (ECM) and form cartilage (see Table 14.1). Matrix components include collagen type II and aggrecan (ACAN).

Long bone formation

In long bone formation hypertrophy of chondrocytes occurs in the periosteal bone collar in the mid-shaft, then osteoclasts reabsorb matrix, cartilage and bone marrow elements and osteoblasts invade to form the primary ossification center that migrates toward the end of the bone shaft. Secondary ossification centers form at the bone-ends. Between the primary and secondary ossification centers there is a cartilaginous region, the growth plate. The growth plate is involved in continuous bone generation until maturity when the primary and secondary ossification centers fuse.

Chondrocyte proliferation is stimulated by growth hormone and insulin-like growth factors IGF1 and IGF2 and by other factors secreted by chondrocytes including Indian hedgehog (IHH) (see Table 14.1).

The progression from chondrocyte proliferation to chondrocyte hypertrophy requires signaling from IHH that binds to the cell surface receptors Patched (PTCH1). This binding relieves SMO (smoothened) inhibition of the GLI1 transcription factors. Fibroblast growth factor receptor FGFR3 plays an important role in suppression of chondrocyte proliferation. This

Table 14.1: Steps in process of cartilage formation and resorption.

Process	Gene products
Chondrocyte differentiation	TRPS1
Chondrocyte proliferation	IGF1, IGF2 IHH, RUNX2, Parathyroid
Chondrocyte hypertrophy	IHH, PTCH1
Suppressor of chondrocyte prolif.	FGF3
Secretion of extra-cellular matrix	Proteoglycans, chondroitin sulfate aggrecan Matrilins, Thrombospondins, COMP, SOX9
Mineralization of matrix	PHOSPHO1, Carminerin1
Chondrocyte death	collagenase, proteinase
Osteoclast 1	SPI, MCSF, BCL2, RANK, TRAF6
Bone resorption	Integrins Cortactin

Note: Symbols and protein names are included in the text above.

importance was demonstrated when activating mutations of FGFR3 were shown to lead to achondroplasia (Shiang *et al.*, 1994). This is the most common non-lethal skeletal dysplasia. It is associated with short-limbed dwarfism, and some degree of limb bowing. In addition, there may be narrowing of the skull foramen magnum and cervical canal. In some cases, cranio–facial abnormalities may be present (Krakow and Rimoin, 2010).

Chondrocytes

Hypertrophy of chondrocytes is important at specific stages of skeletal development and is dependent upon IHH signaling, on expression of parathyroid hormone and on RUNX2 transcription factor. Hydroxyapatite (calcium phosphate) crystals are also deposited in the ECM that surrounds hypertrophic chondrocytes. Mineralization of ECM is also dependent on the activity of phosphatases including alkaline phosphatase and the orphan phosphatase encoded by the *PHOSPHO1* gene. Mineralization is dependent on expression of CARMINERIN1 (CST10), an inhibitor of phosphodiesterase 1 expression.

Chondrocyte secretion of ECM is essential for bone formation. These ECM components include type II collagen fibrils, aggrecan and the glycosaminoglycan hyaluronic acid (Mackie *et al.*, 2011). The *ACAN* gene on chromosome 15 encodes aggrecan. Aggrecan is sometimes referred to as chondroitin sulfate proteoglycan. This protein is heavily modified with glycosaminoglycans and it participates in hyaluronic acid binding and cell adhesion. It is a major component of cartilage and in particular of articular cartilage. Mutations in aggrecan can lead to a skeletal abnormality referred to as osteochondritis dessicans, which involves separation of cartilage and subchondral bone. This condition particularly impacts knees, hips and elbows.

Other important proteins in cartilage related ECM include matrilins (MATN1) and thrombospondin (THBS1). Thrombospondin 5 (THBS5) is also referred to as cartilage oligomeric matrix protein (COMP). Mutations in COMP lead to a condition known as pseudoachondroplasia.

Mackie *et al.* (2011) noted that expression of cartilage ECM proteins is dependent upon the transcription factor SOX9. Mutations in *SOX9* impact development of bone and development of genitalia. *SOX9* mutations lead to Campomelic dysplasia which is a condition associated with bowing of limbs, underdeveloped scapulae, deficiency of ribs and neck abnormalities. Clubbed feet and facial dysmorphology may also occur in Campomelic dysplasia and affected children seldom survive beyond infancy.

Chondrocyte death follows hypertrophy. Degradation of ECM then occurs. Osteoclasts play important roles in ECM degradation. Enzymes involved in degradation include collagenases and cysteine proteinases. Degradation is followed by invasion by bone marrow cells, osteoblasts and vascular elements (see Table 14.1).

Trichorhinophalangeal syndrome arises due to mutations in a gene that encodes a factor that suppresses transcription, regulates proliferation, apoptosis and differentiates chondrocytes. This gene was then named *TRPS1* based on the syndrome that is characterized by skeletal and cranio–facial anomalies and abnormalities of teeth and hair (Momeni *et al.*, 2000).

Osteoclasts

Osteoclasts play essential roles in remodeling and shaping bones during development and throughout life. Mellis *et al.* (2011) reported that osteoclast formation, survival and resorption activities are dependent upon signaling pathways that activate transcription factors.

Osteoclast precursors are hematopoeitic stem cells in the monocyte macrophage lineage. The transcription factor PU1 (SPI1) stimulates macrophage colony stimulating factor (MCSF) and the binding of this factor to its receptors. This binding induces downstream expression of BCL2 and expression of RANK (TNFRF11A) osteoclast receptor and binding of its ligand leads to recruitment of TRAF6 (TNF associated factor). Genes are then activated for osteoclast differentiation.

Mellis *et al.* (2011) reported that osteoclasts have the unique ability to resorb mineralized bone. Impaired osteoclast function leads to a condition known as osteopetrosis. Key to bone resorption are podosomes that form when osteoclasts adhere to bone through integrin adhesion (ITGA and ITGB integrins). Podosomes seal off a compartment in which bone degradation takes place, in part through release of protons and proteases from osteoclasts. SRC kinases play critical roles in podosome aggregation. Mellis *et al.* (2011) emphasized that podosomes must assemble and disassemble in different positions and phosphorylated Cortactin (CTTN) produced by osteoclasts is important in these processes.

Osteoblasts

Osteoblasts are derived from multipotent mesenchymal stem cells. Differentiation of these cells is dependent upon extra-cellular signaling and transcription factor expression. Guntur and Rosen (2011) reported that key signaling pathways involved in osteoblast differentiation include WNT/Beta-catenin and insulin like growth factors (IGF1 and IGF2) and phospho-inositol-3-kinase (PI3K) pathways. Regulation of the PI3K pathway is through the phosphatase PTEN (phosphatase and tensin homolog) that antagonizes PI3K activity.

The PI3K pathway links to the AKT kinase pathway. AKT can phosphorylate transcription factors such as FOXO in the cytoplasm. Transcription factors that play important roles in osteoblast differentiation include RUNX2 that promotes the expression of osterix (SP7) and CREB1 (cyclic AMP responsive element).

Osteoblasts derived from the periosteum form the primary ossification centers of long bones. Terminally differentiated osteoblasts synthesize collagen and additional proteins including osteocalcin and osteopontin, they also produce calcium and phosphate minerals. Osteoblasts trapped in matrix become osteocytes. Osteocytes in mature bone are non-dividing cells.

Interactions between osteocytes, osteoblasts and ECM are critical for bone formation. More than 350 different types of skeletal dysplasia have been described in humans. Osteogenesis imperfecta is a disorder characterized by bone fragility and it is most frequently due to mutations that impact in formation or modification of ECM. Osteogenesis imperfecta is discussed following the section on ECM.

ECM

Approximately 300 proteins constitute the core of the ECM. Hynes and Naba (2012) reviewed the ECM that they referred to as the "matrisome". They noted that biochemical analyses of the ECM are rendered difficult because many of the component proteins are highly insoluble and are cross-linked. Gene studies have revealed that the genes that encode ECM proteins frequently contain many repeated domains.

Frantz *et al.* (2010) emphasized that the ECM varies in different tissues and the different cell types within the tissue, in part, determine these variations. Adhesion occurs between the ECM and receptors on cells, e.g., integrins, syndecans and discoidin receptors. Interactions between the ECM and the cytoskeleton of cells play important roles in cell migration.

Protein elements within the ECM undergo extensive post-translational modifications and these processes are optimized for specific tissue types. The ECM also binds protein such as growth factors and acts as a reservoir of these factors.

Frantz *et al.* (2010) distinguished two main classes of molecular components of ECM, namely proteoglycans and fibrous proteins. Proteoglycans form a gel like substance within the interstitial spaces of the ECM. Key components of proteoglycan are glycosaminoglycans, often linked to a protein core. Small leucine rich proteins often constitute the core. The sugar portion of the glycosaminoglycan comprises polysaccharides, N-acetylglucosamine, D-glucuronic acid, L-glucuronic acid and galactose. In addition the glycosaminoglycans can be sulfated.

Frantz *et al.* (2010) noted that small leucine rich proteoglycans often play roles in signaling processes, either through binding to a receptor or by binding to a ligand of the receptor. Proteoglycans on cell surfaces can act as co-receptors.

Fibrous proteins in the ECM include collagens, elastins, fibronectins, laminins and tenascins. Collagens constitute the most abundant proteins in ECM. Fibroblasts within the interstitial matrix and within tissues secrete collagen and play roles in collagen fiber alignment. Interactions also occur between the different fibrous proteins in the ECM. Collagen and elastin fibers may link; elastin molecules may also crosslink with each other and these crosslinking reactions are often mediated by lysyl oxidases. Unfolded fibronectin molecules expose binding sites that facilitate interactions with the transmembrane receptors on cells such as integrins and syndecans.

Franz *et al.* (2010) reported that there are at least 28 different types of collagen in vertebrates. Collagen molecules assemble into triple stranded helices and these can then interact with other molecules in the ECM. Post-translational processes that collagen undergo include cleavage of propeptides, proline and lysine hydroxylation, and lysine glycosylation. Other important modifying factors are enzymes that promote heparan binding, sulfate binding and interactions with glycosaminoglycans.

Tissue fibroblasts secrete specific forms of collagen, elastin and fibronectin and they also secrete different proteoglycans. Hynes and Naba (2012) considered laminins and fibronectins to be the best-studied ECM glycoproteins. 11 genes in human encode laminins. Fibronectins are encoded by one gene that generates multiple splice isoforms. Tissue homeostasis is also dependent upon secretion of metalloproteinases by fibroblasts.

Basement Membranes

Basement membranes are sheet like ECM structures that adhere to cells. These membranes underlie endothelia and epithelia and play key roles in the structure of many organs. Yurchenco and Patton (2009) reviewed components of basal membranes and their roles in developmental processes. They reported that basement membranes are composed of laminins, nidogens and heparan proteoglycans including agrin and perlecan. Laminins attach the basement membrane to cell surface glycolipids and adhesion is further enhanced through integrins and dystroglycans.

Laminins are trimeric proteins, each composed of variable alpha, beta and gamma chains. Five different genes encode alpha laminins, four different genes encode laminin beta chains and three different genes encode gamma laminin chains. In the laminin molecule, the three different subunits interact to form a cross like structure with linear and globular domains. Specific domains in laminin molecules bind to collagen, integrin or dystroglycan.

Mutations in laminin can lead to pre-implantation lethality. In addition, specific mutations in laminins can lead to abnormalities in muscle, in kidney, vascular endothelial systems and skin. The skin defect can manifest as severe blistering, which in many patients can present in the neonatal period (Tzu and Marinkovich, 2008).

Laminin mutation can lead to cardiomyopathy and arrhythmias (Knoll *et al.*, 2007).

Collagen Processing and Bone Formation, Normal and Abnormal

Insight into processes involved in bone formation has been gained through analysis of conditions associated with increased brittleness of bones, including osteogenesis imperfecta.

Several forms of this genetic disorder are known. Different forms differ in their severity and in the most severe forms bone fractures are present at birth. Osteogenesis imperfecta may be inherited as a genetically dominant condition or as a recessive condition.

There is now evidence that in the majority of cases, there are defects in the biosynthesis and/or structure of collagen, type I collagen or procollagen (Byers and Pyott, 2012). Type I pro-collagen is formed from two chains of pro-alpha 1 encoded by the COL1A1 gene on chromosome 17 and by a single pro-alpha 2 chain encoded by the COL1A2 gene on chromosome 7. The central core of each pro-collagen protein comprises 1,014 amino acids, rich in glycine amino acids.

Byers and Pyott (2012) reported that the post-translational modification of pro-collagen molecules are extensive and require more than 12 distinct proteins. Furthermore, the processes involved in folding and assembly of the proteins to generate extra-cellular fibrils are complex.

Decreased levels of type 1 collagen and in some cases abnormal collagen proteins were identified in dominantly inherited forms of osteogenesis imperfect. Byers and Pyott (2012) reported that recessively inherited osteogenesis imperfecta are most frequently due to mutations in genes that encode products that play roles in processing collagen. Examples include the gene *PLOD2* that encodes lysyl hydroxylase 2. Hydroxylation is an important step in collagen modification. Other important proteins in processing or modification of collagen, include LEPRE1 (a protein hydroxylase) and CRTAP (a cartilage associated protein). Also important in collagen modification and processing are peptidyl-prolyl isomerase (PPIB), CRTAP forms a complex with protein leprecan (LEPR1) and the peptidyl proline hydroxylase also known as cyclophilin.

SERPINH1 and SERPINF1 act as molecular chaperones in collagen biosynthesis. SP7, Osterix (a bone specific transcription factor), functions as an activator for osteoblast differentiation. BMP1 and other bone morphogenic proteins play important roles in endochondral bone formation.

The osteogenesis imperfecta database from Leicester in England http://oi.gene.le.ac.uk lists variants in 14 different genes that can give rise to osteogenesis imperfecta.

CHAPTER 15

KIDNEY AND URINARY TRACT DEVELOPMENT, MALFORMATIONS IN HUMAN GENES

Insight into steps in the development of the kidney and urinary tract systems have been gathered through detailed studies on morphogenesis in the mouse and analysis of genes involved in developmental processes.

Kidney and Urinary Tract Development

Insight into genes involved in human kidney development have been obtained through studies on congenital anomalies of the kidney and urinary tract, and through delineation of specific gene defects that lead to these anomalies. Vivante *et al.* (2014) reviewed kidney and urinary tract anomalies and reported that the most common anomalies include renal agenesis, renal hypoplasia, multiple cystic dysplastic kidney, hydronephrosis, ureter–pelvic junction obstruction, mega-ureter, ureter duplex, vesico–urethral reflux and posterior-urethral valves.

Stages in kidney and urinary tract development

Vivante *et al.* (2014) documented the stages in kidney development. The first stage involved uteric bud induction. The uteric bud arises in the intermediate mesoderm on the dorsal body wall. It invades adjacent mesenchymal tissue, the metanephrogenic mesenchyme. At the tip of the invading bud, mesenchymal to epithelial transformation occurs. Additional bud outgrowth is followed by serial branching with generation of 15 branches. New nephrons are generated at the tip of each branch. Less is known about the subsequent developmental steps that include nephron patterning and establishment of tubular glomerular interactions.

Based on analysis of developmental malformations that occur in association with specific single gene defects gene lists have been constructed for specific developmental stages. The lists are comprehensive for mouse development and are currently less extensive for human development. The first stage, classified as regionalization and uteric bud induction involves at least nine human genes. These include, the growth factor BMP4, ROBO transmembrane receptor, RET receptor tyrosine kinase and transcription factors GATA (zinc finger transcription factor), PAX2, SIX1 and SIX2 (homeobox transcription factors), EYA1 transcriptional co-activator and SALL1 transcriptional repressor.

The developmental stage that includes invasion of the uteric bud into the metanephric mesenchyme and mesenchyme epithelial transition in humans, involves FGF20 and WNT4.

The stage that involves outgrowth and serially branching of the uteric bud and uteric bud morphogenesis involves components of the renin–angiotensin system including renin, encoded by the REN gene, angiotensin converting enzyme (ACE) and angiotensin receptor type 1 (AGTR1).

The stage of segmentation of the nephron and glomerular tubular interaction is less well studied in humans. This stage apparently involves the Uromodulin 1 gene that encodes the Tamm Horsfall protein, the most abundant urinary protein. Kidney defects that arise due to Uromodulin mutations include medullary cystic kidney disease (MCKD), familial juvenile hyperuremic nephropathy (FJHN) and glomerular cystic kidney disease. These are rare diseases that are all inherited as autosomal dominant disorders.

Malformations due to Mutations in Human Genes

Vivante *et al.* (2014) reported that there are single gene defects that lead to renal disease where the development stage at which the defect manifests is not characterized. Important genes in this category include the homeodomain containing transcription factor, hepatocyte nuclear factor 1B (HNF1B) and dual serine threonine and tyrosine kinase regulator (DSTYK) that acts downstream in the fibroblast growth factor (FGF) pathway and is a regulator of ERK (MAPK1) signaling.

HNF1B is reported to be essential for kidney, liver and pancreas embryogenesis. Specific mutations in this gene lead to renal cysts diabetes syndrome (RCDS). Valente *et al.* (2014) reported that mutation in this gene also led to renal hypoplasia, multi-cystic dysplastic kidney, cystic kidney disease and single kidney. The renal diseases caused are characterized by hyper-uricemia and hypo-magnesemia. Based on studies in mice, there is evidence that HNF1B stimulates gene transcription of the *PKHD1* gene (autosomal recessive polycystic kidney disease).

Vivante *et al.* (2014) reported that the most common defects encountered in congenital anomalies of the kidney and urinary tract impact the transcription factors PAX2 and HNF1B.

It is interesting to note that single gene defects that lead to congenital anomalies of the kidney and urinary tract are also frequently associated with extra-renal anomalies. These frequently involve the ear and include deafness and sometimes branchial cysts (located in the neck) occur with mutations in any one of the following genes: *EYA1*, *GATA*, *SIX1*, *SIX5* and *SALL1*. Eye abnormalities may be present in cases with abnormalities in BMP4 and PAX2.

Gimelli *et al.* (2010) identified a duplicated segment of chromosome 8 in a young female with congenital abnormalities of the kidney and urinary tract. One of the genes present in this duplication was *SOX17*. They subsequently identified specific mutations in the SOX17 transcription factor in five other patients with congenital abnormalities of the kidney and urinary tract. Their studies revealed that the mutant SOX17 protein accumulated in cells and interfered with normal functioning of the WNT signaling pathway.

CHAPTER 16

SEX DETERMINATION DIFFERENTIATION
AND ENDOCRINE GLANDS

Chromosomal, Gonadal and Phenotypic Sex

Sex is determined by the genomic constitution: XX in females and XY in males. Sex differentiation is dependent on factors produced in embryonic gonads. Despite comprehensive studies in vertebrates over several decades precise molecular mechanisms involved in determining sexual differentiation have not yet been elucidated. Insights into factors involved in sexual differentiation have in part been obtained through studies in patients with disorders of sexual development. These are defined as congenital conditions where there are discrepancies between chromosomal, gonadal and phenotypic sex. Baxter and Vilain (2013) reported that despite advances, in many patients with disorders of sexual differentiation, a definitive genetic diagnosis was still not possible.

In the early human embryo by the 5th week post-conception primordial germ cells have migrated from the yolk sac to the mesenchyme of the posterior body. These cells then move to a position medial and ventral to the mesonephros (Larsen, 2009). Cells from the coelomic epithelium surround the germ cells and act as support cells for the undifferentiated gonad.

By the 6th week para-mesonephric ducts, the Mullerian ducts, are distinguishable in both male and female embryos. The Mullerian duct is bilateral in a position lateral to the mesonephros and the mesonephric duct. The caudal ends of the Mullerian duct fuse.

Testes development

At approximately the 6th week stage male genital differentiation commences. The signal for this development is expression of the *SRY* gene on the Y chromosome. A key function of *SRY* is to increase transcription of the *SOX9* gene. SRY expression is the key to differentiation of Sertoli cells from the support cells. There is evidence that Sertoli cell differentiation is critical to the subsequent differentiation of Leydig cells and hormone production. Sertoli cells also promote differentiation of the cords that enclose the germ cells in the male gonad and Sertoli cells promote formation of the testis specific vascular system.

Other key gene products important in male sex determination include SOX3, a protein that is closely related to *SRY*, and the transcription factor GATA2 and its cofactor FOG2 (friend of GATA2) also impact testis development. There is evidence that normal expression of GATA2 is necessary for production of anti-Mullerian factor (Baxter and Vilain, 2013).

It is important to note that the discussion here includes only gene products that have been shown to impact human sex determination. Additional gene products have been shown to be important in testis determination in mouse but at the time of writing their roles in humans are not yet confirmed.

Ovarian development

In the XX gonad, the somatic support cells derived from coelomic epithelium differentiate into granulosa cells that surround the germ cells and the germ cells go on to produce oogonia. The mesonephric duct degenerates and the Mullerian duct develops further (Larsen, 2009).

Key factors involved in differentiation of the ovary include WNT4 signaling under the influence of RSPO1 and the production of the transcription factor FOXL2 (Biason-Lauber, 2012). WNT4 signaling is detectable in the granulosa cells and in the oocytes. WNT4 and RSPO1 act together to promote synthesis and stabilization of beta-catenin. Beta-catenin acts to inhibit the expression of SOX9 and it therefore has an anti-testis effect.

In addition, WNT4 signaling up-regulates expression of DAX1 (NROB1 nuclear receptor) that inhibits enzymes involved in testosterone synthesis. Deletion of WNT4 leads 46XX females to undergo females to male sex reversal. Excessive WNT4 through gene duplication can lead XY fetuses to undergo male to female sex reversal (Biason-Lauber, 2012).

RSPO is produced by keratinocytes and is also necessary for normal ovary development (Chassot *et al.*, 2008). Mutations leading to deficiency of RSPO1 have been found in 46XX *SRY* negative hermaphrodites who have both male and female gonadal elements (Tomaselli *et al.*, 2008).

Expression of the transcription factor FOXL2 is one of the earliest signals of ovarian differentiation. FOXL2 represses SOX9 production. FOXL2 is required for normal development of the ovary and the eyelids. Defects in FOXL2 lead to an autosomal dominant disorder associated with eyelid abnormalities and premature ovarian failure. FOXL2 is also expressed in the pituitary in gonadotropic and thyrotrophic cells. In the pituitary, it plays a role in stimulating expression of the glycoprotein hormone subunit alpha-GSU that forms a subunit common to follicle stimulating hormone (FSH), luteinizing stimulating hormone (LSH) and thyroid stimulating hormone (TSH).

An additional transcription factor involved in gonadal development is NRSA1 (SF1), steroidogenic factor that is important for development of the testis and adrenal gland. It is however also present in the ovary in granulosa cells.

There is also evidence that MAP3K1 signaling plays a role in normal sexual development. MAP3K1 mutations were identified in cases of XY gonadal dysgenesis (Baxter and Vilain, 2013).

Female Protective Effect

In a number of diseases, e.g., the neurocognitive disease autism, there is a much higher frequency of the disease in males than in females. There are some concerns that diagnostic bias may be a factor in defining a disease as more common in one gender than the other, nevertheless it is clear for autism that there is an increased frequency in males.

Jeste and Geschwind (2014) discussed two possible explanations for increased male frequency of autism. The first has to do with a "protective effect" from the second X chromosome, the "female protective effect". Support for this comes from the observation that the incidence of autism is higher in XO Turner syndrome individuals than in XX females and autism frequency is higher in XYY males than in XY males. Another possible cause of the higher autism frequency in males may be the effect of testosterone on brain development. Baron-Cohen *et al.* (2014) accessed stored amniotic fluid samples that were collected between 1993 and 1999. They determined levels of steroidogenic hormones in these samples and also accessed diagnostic records to determine which samples originated from individuals who subsequently developed autism. Their studies revealed that testosterone levels were higher in amniotic fluids obtained from children who subsequently went on to develop autism or Asperger syndrome.

Endocrine Gland Development

Thyroid development

In a review of thyroid development, normal and abnormal, Szinnai (2014) reported that the human thyroid gland originates from two lateral primordial sources and one median primordial source. The lateral primordia are neuroectodermal in origin and are derived from the 4th pharyngeal arch. The median primordial cells originate from foregut endoderm. The connection of the lateral thyroid primordia to the pharyngeal arch is disrupted by day 32. These primordia then descend and attach to the thyroglossal duct to a position below the larynx. The thyroglossal duct subsequently breaks down. Following migration to the pre-tracheal position, fusion of the median and the lateral thyroid anlagen occurs and differentiation of the thyroid takes place.

Key transcription factors involved in the early stages of thyroid development in humans include NKX2-1, PAX8 and FOXE1. Szinnai (2014) reported that terminal differentiation of the thyroid commences by day 48; this involves structural and functional changes. Thyroglobulin (TG) is produced and thyroid follicle cells differentiate. These cells subsequently adhere to each other to form thyroid follicles. The appearance of small follicles is referred to as the colloid stage and at this stage, the thyroid begins to concentrate iodine and thyroid hormone synthesis begins.

Iodine uptake is dependent on the action of solute carriers SLC26A and NIS. Specific proteins essential for thyroid hormone synthesis include thyroid peroxidase (TPO), TG, dual oxidase (DUOX), TSH receptor (TSHR).

Transcription factor mutations and thyroid dysgenesis

Specific transcription factor mutations occur in specific syndromes in which thyroid dysgenesis occurs as one of the manifestations. FOXE1 mutations occur in Bamforth syndrome that is characterized by athyreosis (absent thyroid), cleft palate and spiky hair. NKX2-1 mutations occur in brain lung thyroid syndrome. NKX2-1 is essential for production of surfactant protein. PAX8 mutations lead to a syndrome associated with thyroid dysgenesis and kidney abnormalities.

Non-syndromic disorders, associated primarily with thyroid disease also occur. Mutation in the TSHR may lead to hypoplastic thyroid gland and hypothyroidism. The frequency of disease causing mutation in TSHR varies in different populations but on an average account for 4.3% of cases of congenital hypothyroidism. PAX8 inactivating mutations have also been found in patients with hypothyroidism. Szinnai (2014) noted however that thyroid dysfunction is most likely multigenic in origin and due to the combined effects of mutations in a number of different genes. It is also clear that environmental factors including iodine availability impact thyroid hormone production.

Pendred syndrome

Pendred syndrome is an autosomal recessive condition associated with defects in iodine transport and processing in the thyroid. Pendred syndrome occurs in cases with bi-allelic mutations in the iodine solute carrier SLC26A. In this syndrome there is enlargement of the thyroid, hypothyroidism and pre-lingual deafness, in the inner ear there is enlargement of the endolymphatic system with enlargement of the vestibular aqueduct (Kopp, 2014). However, the phenotypic spectrum varies. Thyroid abnormalities in patients with SLC26A mutations may include hypoplastic thyroid and thyroid enlargement (goiter).

TPO mutations have been described in cases of congenital hypothyroidism. Clinical manifestations include thyroid enlargement (goiter). Recessive TPO mutations are not uncommon causes of hypothyroidism in Israeli and Arab populations (Tenenbaum-Rakover *et al.*, 2007). Mutations in the *DUOX2* gene may also impair iodination and thyroid hormone synthesis. *DUOX2* gene encodes an enzyme that generates peroxide; it also transports iodide. The TPO requires hydrogen peroxide to carry out its function.

Pituitary Gland Development and Function

The ectoderm between the maxillary and mandibular processes (the sto-modeum) invaginates to give rise to Rathke's pouch, during the 3rd week post-conception. This grows toward a diverticulum, the infundibulum that develops from the floor of the diencephalon. Rathke's pouch subsequently separates from the stomodeum and differentiates to form the anterior, and intermediate pituitary lobes, the adenohypophysis. The posterior lobe of the pituitary is derived from the infundibulum and the neuroectoderm of the diencephalon and forms the neurohypophysis (Larsen, 2009). Signaling molecules important in pituitary development include BMP4 that is expressed by cells in the diencephalic floor and FGF8 expressed in the infundibulum.

Kelberman and Dattani (2007) noted that inductive signals from the neuroectoderm are important for initial development of the pituitary. In addition, a cascade of signaling molecules and transcription factors play roles in subsequent pituitary development and differentiation of its cell types. They emphasized that though much is known about development of the pituitary in the rodent, less is known about human pituitary development. However a number of genes that encode specific transcription factors have been found to be mutated in particular cases of hypopituitarism in humans. These include genes found to harbor mutations in syndromic forms of hypopituitarism including transcription factors HESX1, LHX3, LHX4, SOX2, SOX3, TBX19 and ARNT2 (aryl hydrocarbon nuclear receptor). In syndromic forms of hypopituitarism, additional malformations occur and may include abnormalities of the optic nerve and eye, defects in the septum pellucidum and corpus callosum, short neck and sensorineural hearing loss. Non-syndromic forms of hypopituitarism may arise due to defects in the transcription factors POU1F1 (PIT1) and PROP1; in these cases multiple pituitary hormones may be deficient.

LHX3 and LHX4 are LIM homeobox transcription factors. LHX3 is strongly expressed in Rathke's pouch and is then expressed in the pituitary throughout development and also in adult life. Homozygous deletions of LHX3 have been reported in families with anterior pituitary hormone deficiency. LHX4 is closely related to LHX3 and is expressed primarily in Rathke's pouch.

HESX1 mutations were identified in patients with septo-optic dysplasia, a congenital dysplasia associated with optic nerve abnormalities, brain midline defects, including corpus callosum and septum pellucidum defects and pituitary hypoplasia. The range of defects differed in different patients and some patients also manifested seizure disorders. It is however important to note that septo-optic dysplasia also occurs in patients without HESX1 defects and it is likely a multi-factorial disorder in which environment also plays roles.

PROP1 is a transcription factor expressed almost exclusively in the embryonic pituitary. Recessive mutations in PROP1 have been reported in cases with deficiency of growth hormone, TSH and gonadotrophins. The transcription factor POU1F (PIT1) is expressed late in pituitary development and throughout life. Kelberman and Dattani (2007) reported that it is essential for somatotrophic, lactotrophic and thyrotrophic lineages.

The adenohypophysis includes the anterior and intermediate lobes of the pituitary. Distinct cell types in the adenohypophysis produce six different hormones. Somatotropes secrete growth hormones, lactotropes produce prolactin, corticotropes produce Adrenocorticotropic hormone (ACTH), thyrotropes produce TSH, gonadotropes produce FSH and LSH.

Hypothalamic corticotrophin releasing hormone activates transcription of pro-opiomelanocortin (POMC). The intermediate lobe secretes POMC that is cleaved to give rise to ACTH and melanocyte stimulating hormone (MSH).

Axonal terminals of the magnocellular neurons located in the hypothalamus are the main constituents of the posterior lobe. The hypothalamic magnocellular neurons produce oxytocin and vasopressin.

Hypothalamic releasing hormones regulate the release of hormones from the pituitary. These include corticotrophin releasing hormone and thyroid releasing hormone synthesized by the paraventricular nucleus in the hypothalamus. Gonadotrophin releasing hormone is secreted by the hypothalamic arcuate nucleus and the ventromedial nucleus. The hypothalamus also produces growth hormone releasing factor.

There is evidence that the patterning and induction of secretion in the hypothalamus is regulated by signaling molecules Hedgehog (HH), Nodal, BMP (bone-morphogenic protein) and WNT gene products (Zhu *et al.*, 2007).

Adrenal Gland Development and Defects in Function

In the human fetus the adrenal anlagen arise in a position posterior and medial to the urogenital ridge at approximately 33 day post-conception (Ishimoto and Jaffe, 2011).

The transcription factors SF1 and DAX1 (also known as NROB1) are expressed in the adrenal anlage. By day 50, distinct zones are distinguishable in the primitive adrenal gland; these include an inner zone, the fetal zone with large eosinophilic cells and an outer definitive zone where small basophilic cells occur. Subsequently, a transitional zone develops between these inner fetal and the outer definitive zones. These zones constitute the early adrenal cortex. More detailed information is available on steps in the development of the adrenal cortex than is available on development of the adrenal medulla. However, Inoue *et al.* (2010) identified neural crest derived chromaffin positive ganglion cells and Schwann cells in fetuses between 9 and 16 weeks of gestation. These cells ultimately form the adrenal medulla. They reported that tyrosine hydroxylase positive chromaffin cells entered the adrenal gland at different points. These cells develop further to produce catecholamines, epinephrine and nor-epinephrine.

Insights into signaling cascades and transcription factors that play roles in the adrenal cortex have been obtained through studies in patients with congenital adrenal disorders and in studies on mouse models of adrenal disease where specific genes were knocked out (Kempna and Fluck, 2008). Adrenal hypoplasia or absence of adrenal glands was reported in patients with mutations in the GLI3, a key factor in the HH signaling pathway. Abnormalities of adrenal function have also been reported in patients with mutations in *WT1* (Wilm's tumor1), *WNT4* and *SALL1* genes. Other important factors in adrenal development include FOXD2 and PBX1.

As noted above SF1 and DAX1 are key factors in adrenal development from early on. DAX1 maps to the X chromosome. Kempna and Fluck (2008) reported that males with DAX1 mutations may present with hypogonadism and adrenal insufficiency. Female carriers (i.e., heterozygotes) of DAX1 mutations are normal. The *NR5A1* gene on chromosome 9q33 encodes the SF1. Bi-allelic mutations in the *NR5A1* gene were first described in patients with adrenal insufficiency and disorders of sexual development. NR5A mutations have also been reported in patients with primary ovarian insufficiency.

Hypothalamic and pituitary defects leading to defects in synthesis and processing of POMC may lead to adrenocortical hypoplasia. POMC is processed to produce ACTH, MSH and beta-endorphin. This processing requires the enzyme pro-hormone convertase 1 (PC1) Mutations in this gene lead to adrenal cortical hypoplasia and adrenal insufficiency. Kempna and Fluck (2008) noted that ACTH deficiency and adrenal hypoplasia occur as features of pituitary functional defects that result from mutations in transcription factors involved in pituitary development which include HESX1, LHX4, SOX3 and PROP1.

Mutations in genes encoding the ACTH receptor (MC2R) may also lead to adrenal insufficiency.

Ishimoto and Jaffe (2011) reported that the adrenal gland is one of the most highly vascularized organs in the human fetus. This angiogenesis is mediated by fibroblast growth factor (FGF) and vascular endothelial growth factor (VEGF).

Genes involved in adrenal synthesis of steroid hormones

Key factors involved in steroid hormone synthesis in the adrenal gland include STAR (steroid acute regulatory protein) and the cytochrome p450 enzymes involved in conversion of cholesterol to steroid hormones. The first reaction in this pathway involves CYP11A that converts cholesterol to pregnenolone.

Decreased synthesis of cortisol results from a number of different defects in the pathway of steroid biogenesis. Adrenal hyperplasia results from increased synthesis of pituitary ACTH in consequence of the limited cortisol.

Deficiency of 21 hydroxylase

Deficiency of 21 hydroxylase, the enzyme encoded by the *CYP21A2* gene is the commonest cause of the disorder. The birth incidence of this disorder varies in different populations and reaches highest frequency in the Eskimo population (1 in 500 to 1 in 1,500). The *CYP21A2* gene and the closely related pseudogene *CYP21A1P* map in the HLAIII region on chromosome 6p13.3. These two genes are 98% identical; however the pseudogene *CYP21A1P* is not expressed due to deletions in the gene. In some cases exchange of sequences between *CYP21A1P* and *CYP21A2* occurs, leading to gene disruption (gene conversion). The *CYP21A2* encoded product is dependent on NADP for its activity and on cytochrome p450 oxidoreductase, encoded by the *POR* gene (see Figure 16.1).

Krone and Arlt (2009) reported that inactivating *CYP21A2* mutations include mutations, gene conversion and deletions; an 8bp deletion in exon 3 is relatively common, other rare mutations also occur. Since *CYP21A2* has 10 exons and short introns, Krone and Arlt (2009) noted that

21-Hydroxylase reactions

Progesterone-----➜- Deoxycorticosterone

17-Alpha hydroxyprogesterone--➜-11 Deoxycorticosterone

21-hydoxylase deficiency leads to

Deficiency of deoxycorticosterone

Deficiency of 11-deoxycorticosterone

Deficiency of aldosterone derived from corticosterone

21-hydoxylase deficiency leads to

Raised levels of Progesterone and 17-Alpha hydroxyprogesterone

17-Alpha hydroxyprogesterone---Androstendione

Androstenedione ----Testosterone----
Dihydrotestosterone

Fig. 16.1:   21 Hydroxylase deficiency.

sequencing of the entire gene is the best way to identify all mutations. They noted that 65–85% of patients have compound heterozygous mutations.

Phenotypes in 21 hydroxylase deficiency

The classical form of this deficiency presents in the newborn period with a salt wasting crisis and/or virilization of female genitalia due to accumulation of 17-OH progesterone and its conversion to androstenedione and testosterone (New MI, 2004). Non-classical forms may present later in life and these presentations include precocious puberty and females may present with polycystic ovary syndrome and hirsutism.

Function of 21 hydroxylase

The enzyme 21 hydroxylase functions in the conversion of progesterone to deoxycorticosterone and conversion of 17 alpha-hydroxyprogesterone to 11-deoxycortisol. Deoxycorticosterone is then converted to corticosterone through the activity of 11 beta-hydroxylase and that enzyme also converts 11-deoxycortisol to cortisol.

In the absence of 21 hydroxylase, progesterone accumulates and enters pathways to adrenal androgen synthesis leading to increased levels of androstenedione and testosterone.

11 beta-hydroxylase deficiency: CYP11B1 mutations

Congenital adrenal hyperplasia may also occur due to impaired function of the enzyme 11 beta-hydroxylase. In the absence of this enzyme, there is inadequate synthesis of corticosterone, glucocorticosteroids and aldosterone. Krone and Arlt (2009) reported that an accumulation of deoxycorticosterone in this disorder activates the mineralocorticoid receptor leading to hypertension.

Deficiency of 11 beta-hydroxylase has a high incidence in Israel and Morocco than in other populations.

The *CYP11B1* gene lies within 40 kb of the homologous *CYP11B2* gene. In rare cases, crossing over between *CYP11B1* and *CYP11B2* occur leading to deficiency and disease.

17 alpha-hydroxylase deficiencies due to CYP17A1 mutations

Krone and Arlt (2009) reported that approximately 1% of cases of congenital adrenal hyperplasia are due to defect in the function of 17 alpha-hydroxylase. Two different types of reactions are catalyzed by 17 alpha-hydroxylase. The first is the hydroxylation reaction. Hydroxylation converts pregnenolone to 17 alpha-hydroxyl-pregnenolone and progesterone to 17 alpha-progesterone. The second reaction carried out by 17 alpha-hydroxylase is a lyase reaction. Lyase reactions convert 17 alpha-hydroxy pregnenolone to dehydroepiandrosterone (DHEA) and 17 alpha-hydroxyl-progesterone to androstendione.

Defects in 17 alpha-hydroxylase lead to deficiency in glucocorticoids. In some cases, the lyase function of the enzyme is primarily impaired and delays puberty results.

Congenital adrenal hyperplasia due to deficiency of 3 beta-hydroxy steroid dehydrogenase (HSD3B2)

3 beta-hydroxy steroid dehydrogenase catalyzes conversion of pregnenolone to progesterone; conversion of 17 alpha-pregnenolone to 17 alpha-progesterone; conversion of DHEA to androstenedione and conversion of androstenediol to testosterone.

Since conversion of pregnenolone to progesterone is limited, it impacts the generations of corticosterone and aldosterone. This, therefore leads to a variety of clinical manifestations including salt wasting and ambiguous genitalia.

Congenital adrenal hyperplasia due to deficiency of cytochrome p450 NADPH oxidoreductase

This enzyme protein is located in the endoplasmic reticulum and it supplies electrons from reduced NADP for the catalytic activity of enzymes involved in steroid biosynthesis. It is also involved in many other metabolic reactions. This enzyme is encoded by the *POR* gene located on chromosome 17q11.2. It has 15 protein coding exons and encodes NADP and flavin binding domains that bind flavin adenine dinucleotide

(FAD) and FMN (flavin mononucleotide). One untranslated exon is also present in the gene.

Founder mutations occur in this gene in the Japanese population. In cases with *POR* mutations, the congenital adrenal hyperplasia related symptoms are mild. Additional unusual features occur including skeletal defects.

CONCLUSION

There are still large caveats in our understanding of birth defects. It is important to consider newer explorations and concepts to expand our knowledge. More information will need to be gathered to understand cell specific and contextual differences in response to specific signaling factors.

In a study of transforming growth factor signaling, Massague and Xi (2012) emphasized a number of contextual differences that determine response. These included ligand isoform differences, ligand blockers, crosstalk between downstream effectors of signaling, chromatin status, DNA binding of transcription factors and cofactors that enhance or repress binding.

In a study on digit formation in mice, Rospovic *et al.* (2014) reported that there is a network of diffusible morphogens that impact the period expression of the transcription factor SOX9. The positive morphogen, bone morphogenetic protein (BMP) and the inhibitory morphogen WNT determine periodicity of expression of *Sox9*. BMP inhibition led to a loss of digits, whereas WNT inhibition led to a loss of interdigital spaces.

There is growing emphasis on transcription factor networks and on the combinatorial cross regulation of transcription factors (Neph *et al.*, 2012).

Population studies, family studies and DNA sequence studies indicate that many congenital defects are not monogenic or oligogenic disorders but are polygenic in nature. Interactions between mutations or variants in different genes are likely causative. The variants may be rare or they may be not uncommon in a specific population (Grosen *et al.*, 2010a). Furthermore, interactions between gene variants and environmental factors must be considered. A gene alteration may act as a risk factor and the full manifestation of the defect may be dependent on environmental factors.

Facilitated sequencing and the ability to sequence DNA from affected tissues and unaffected tissues have revealed that somatic mosaicism is important in a number of different genetic disorders associated with structural malformations. Mosaicism most commonly results from post-zygotic mutations or post-zygotic chromosome changes. One example is enlargement of one part of the brain, hemimegancephaly that may be due to somatic mosaicism involving mutations in genes *PIKCA*, *AKT* or *MTOR* (Poduri *et al.*, 2013). Another example is CLOVES syndrome (Kurek *et al.*, 2012). This syndrome is associated with congenital lipomatous overgrowth, frequently of one limb, associated with vascular, epidermal and skeletal anomalies. It is due to post-zygotic mutations in the *PIK3CA* gene that encodes the catalytic subunit of the phosphatidylinositol-3 kinase.

REFERENCES

A

Abbott NJ, Patabendige AA, Dolman DE, Yusof SR and Begley DJ. Structure and function of the blood-brain barrier. *Neurobiol Dis.* **2010** Jan; 37(1):13–25. doi: 10.1016/j.nbd.2009.07.030. Epub 2009 Aug 5.

Ackerman KG and Pober BR. Congenital diaphragmatic hernia and pulmonary hypoplasia: new insights from developmental biology and genetics. *Am J Med Genet C Semin Med Genet.* **2007** May 15; 145C(2):105–108. PMID: 17436306.

Aguirre A, Dupree JL, Mangin JM and Gallo V. A functional role for EGFR signaling in myelination and remyelination. *Nat Neuro sci.* **2007** Aug; 10(8):990–1002. Epub 2007 Jul 8. PMID:17618276.

Ali FR, Cheng K, Kirwan P, Metcalfe S, Livesey FJ, Barker RA and Philpott A. The phosphorylation status of Ascl1 is a key determinant of neuronal differentiation and maturation *in vivo* and *in vitro*. Development. **2014** Jun; 141(11): 2216–2224. doi: 10.1242/dev.106377. Epub 2014 May 12. PMID: 24821983.

Al-Hussaini A, Faqeih E, El-Hattab AW, Alfadhel M, Asery A, Alsaleem B, Bakhsh E, Ali A, Alasmari A, Lone K, Nahari A, Eyaid W, Al-Balwi M, Craig K, Butterworth A, He L and Taylor RW. Clinical and molecular characteristics of mitochondrial DNA depletion syndrome associated with neonatal cholestasis and liver failure. J Pediatr. **2014** Mar; 164(3):553–559. e1–2. doi: 10.1016/j.jpeds.2013.10.082. Epub 2013 Dec 8. PMID: 2432153.

Amorim MR, Lima MA, Castilla EE and Orioli IM. Non-Latin European descent could be a requirement for association of NTDs and MTHFR variant 677C > T: a meta-analysis. Am J Med Genet A. **2007** Aug 1; 143A(15):1726–1732. PMID:17618486.

Amyere M, Aerts V, Brouillard P, McIntyre BA, Duhoux FP, Wassef M, Enjolras O, Mulliken JB, Devuyst O, Antoine-Poirel H, Boon LM and Vikkula M. Somatic uniparental isodisomy explains multifocality of glomuvenous malformations. Am J Hum Genet. **2013** Feb 7; 92(2):188–196. doi: 10.1016/j.ajhg.2012.12.017. Epub 2013 Jan 31. PMID: 23375657.

Andersen TA, Troelsen-Kde L and Larsen LA. Of mice and men: molecular genetics of congenital heart disease. Cell Mol Life Sci. **2014** Apr; 71(8): 1327–1352. doi:10.1007/s00018-013-1430-1. Epub 2013 Aug 10. PMID: 23934094.

Arcondéguy T, Lacazette E, Millevoi S, Prats H and Touriol C. VEGF-A mRNA processing, stability and translation: a paradigm for intricate regulation of gene expression at the post-transcriptional level. Nucleic Acids Res. **2013** Sep; 41(17):7997–8010.doi:10.1093/nar/gkt539. Epub 2013 Jul 12. PMID: 2381566.

Arnaud P. Genomic imprinting in germ cells: imprints are under control. Reproduction. **2010** Sep; 140(3):411–423. doi:10.1530/REP-10-0173. Epub 2010 May 25. PMID: 20501788.

Autry AE and Monteggia LM. Brain-derived neurotrophic factor and neuropsychiatric disorders. Pharmacol Rev. **2012** Apr; 64(2):238–258. doi: 10.1124/pr.111.005108. Epub 2012 Mar 8. Review. PMID: 22407616.

Avrahami D and Kaestner KH. A cistrome roadmap for underst and ing pancreatic islet biology. Nat Genet. **2014** Feb; 46(2):95–96. doi: 10.1038/ng.2880. PMID: 24473321.

B

Babiarz JE, Ruby JG, Wang Y, Bartel DP, Blelloch R. Mouse ES cells express endogenous shRNAs, siRNAs, and other Microprocessor-independent, Dicer-dependent small RNAs. Genes Dev. **2008** Oct 15; 22(20): 2773–2785. doi: 10.1101/gad.1705308. PMID:18923076.

Baerg J, Kaban G, Tonita J, Pahwa P and Reid D. Gastroschisis: a sixteen-year review. J Pediatr Surg. **2003** May; 38(5):771–774. PMID: 12720191.

Baron MH, Isern J and Fraser ST. The embryonic origins of erythropoiesis in mammals. Blood. **2012** May 24; 119(21):4828–4837. doi: 10.1182/blood-2012-01-153486. Epub 2012 Feb 15. Review. PMID: 22337720.

Baron-Cohen S, Auyeung B, Nørgaard-Pedersen B, Hougaard DM, Abdallah MW, Melgaard L, Cohen AS, Chakrabarti B, Ruta L and Lombardo MV. Elevated fetal steroidogenic activity in autism. Mol Psychiatry. **2014** Jun 3. doi: 10.1038/mp.2014. 48. [Epub ahead of print]. PMID: 24888361.

Barros CS, Franco SJ and Müller U. Extra-cellular matrix functions in the nervous system. Cold Spring Harb. Perspect. Biol. **2010**, 3, a005108. doi: 10. 1101/cshperspect.a005108 publ.2011.

Baskin B, Choufani S, Chen YA, Shuman C, Parkinson N, Lemyre E, Micheil-Innes A, Stavropoulos DJ, Ray PN and Weksberg R. High frequency of copy number variations (CNVs) in the chromosome 11p15 region in patients with Beckwith–Wiedemann syndrome. Hum Genet. **2014** Mar; 133(3):321–330. doi: 10.1007/s00439-013-1379-z. Epub 2013 Oct 24. PMID: 24154661.

Bauer DE, Kamran SC, Lessard S, Xu J, Fujiwara Y, Lin C, Shao Z, Canver MC, Smith EC, Pinello L, Sabo PJ, Vierstra J, Voit RA, Yuan GC, Porteus MH,

Stamatoyannopoulos JA, Lettre G and Orkin SH. An erythroid enhancer of BCL11A subject to genetic variation determines fetal hemoglobin level. Science. **2013** Oct 11; 342(6155):253–257. doi: 10.1126/science.1242088. PMID: 24115442.

Baxter RM and Vilain E. Translational genetics for diagnosis of human disorders of sex development. Annu Rev Genomics Hum Genet. **2013**; 14:371–392. doi: 10.1146/annurev-genom-091212-153417. Epub 2013 Jul 15. PMID: 23875799.

Bell KN and Oakley GP Jr. Update on prevention of folic acid-preventable spina bifida and anencephaly. Birth Defects Res A Clin Mol Teratol. **2009** Jan; 85(1):102–107. doi: 10.1002/bdra.20504. PMID: 19067404.

Bernstein BE, Mikkelsen TS, Xie X, Kamal M, Huebert DJ, Cuff J, Fry B, Meissner A, Wernig M, Plath K, Jaenisch R, Wagschal A, Feil R, Schreiber SL and Lander ES. A bivalent chromatin structure marks key developmental genes in embryonic stem cells. Cell. **2006** Apr 21; 125(2):315–326. PMID: 16630819.

Berry GT. Galactosemia: when is it a newborn screening emergency? Mol Genet Metab. **2012** May; 106(1):7–11. doi: 10.1016/j.ymgme.2012. 03. 007. Epub 2012 Mar 21. Review. PMID: 22483615.

Beurskens LW, Tibboel D, Lindemans J, Duvekot JJ, Cohen-Overbeek TE, Veenma DC, de Klein A, Greer JJ and Steegers-Theunissen RP. Retinol status of newborn infants is associated with congenital diaphragmatic hernia. Pediatrics. **2010** Oct; 126(4):712–720. doi: 10.1542/peds.2010.

Benitez CM, Goodyer WR and Kim SK. Deconstructing pancreas developmental biology. Cold Spring Harb Perspect Biol. **2012** Jun 1; 4(6): a012401. doi: 10.1101/cshperspect.a012401. PMID: 22587935.

Beygo J, Ammerpohl O, Gritzan D, Heitmann M, Rademacher K, Richter J, Caliebe A, Siebert R, Horsthemke B and Buiting K. Deep bisulfite sequencing of aberrantly methylated loci in a patient with multiple methylation defects. PLoS One. **2013** Oct 9; 8(10):e76953. doi: 10.1371/journal.pone. 0076953.eCollection2013.

Biason-Lauber A. WNT4, RSPO1, and FOXL2 in sex development. Semin Reprod Med. **2012** Oct; 30(5):387–395. doi: 10.1055/s-0032-1324722. Epub 2012 Oct 8. Review. PMID: 23044875.

Bickmore WA. The spatial organization of the human genome. Annu Rev Genomics Hum Genet. **2013**; 14:67–84. doi: 10.1146/annurev-genom-091212-153515. Epub 2013 Jul 15. PMID: 23875797.

Bisschoff IJ, Zeschnigk C, Horn D, Wellek B, Rieß A, Wessels M, Willems P, Jensen P, Busche A, Bekkebraten J, Chopra M, Hove HD, Evers C, Heimdal K,

Kaiser AS, Kunstmann E, Robinson KL, Linné M, Martin P, McGrath J, Pradel W, Prescott KE, Roesler B, Rudolf G, Siebers-Renelt U, Tyshchenko N, Wieczorek D, Wolff G, Dobyns WB and Morris-Rosendahl DJ. Novel mutations including deletions of the entire OFD1 gene in 30 families with type 1 orofaciodigital syndrome: a study of the extensive clinical variability. Hum Mutat. **2013** Jan; 34(1):237–247. doi: 10.1002/humu.22224. Epub 2012 Oct 17. PMID: 23033313.

Bizet AA, Tran-Khanh N, Saksena A, Liu K, Buschmann MD and Philip A. CD109-mediated degradation of TGF-β receptors and inhibition of TGF-β responses involve regulation of SMAD7 and Smurf2 localization and function. J Cell Biochem. **2012** Jan; 113(1):238-46. doi: 10.1002/jcb.23349. PMID: 21898545.

Bondurand N, Fouquet V, Baral V, Lecerf L, Loundon N, Goossens M, Duriez B, Labrune P and Pingault V. Alu-mediated deletion of SOX10 regulatory elements in Waardenburg syndrome type 4. Eur J Hum Genet. 2012 Sep; 20(9):990–994. doi: 10.1038/ejhg.2012. 29. Epub **2012** Feb 29. PMID: 22378281.

Brems H, Pasmant E, Van Minkelen R, Wimmer K, Upadhyaya M, Legius E and Messiaen L. Review and update of SPRED1 mutations causing Legius syndrome. Hum Mutat. **2012** Nov; 33(11):1538–1546. doi: 10.1002/humu.22152. Epub 2012 Aug 1. PMID: 22753041.

Brent RL. Environmental causes of human congenital malformations: the pediatrician's role in dealing with these complex clinical problems caused by a multiplicity of environmental and genetic factors. Pediatrics. **2004** Apr; 113 (4 Suppl):957–968. Review. PMID: 15060188.

Brouillard P and Vikkula M. Genetic causes of vascular malformations. Hum Mol Genet. **2007** Oct 15; 16(2):R140–149. Epub 2007 Jul 31. PMID:17670762

Brunetti-Pierri N, Berg JS, Scaglia F, Belmont J, Bacino CA, Sahoo T, Lalani SR, Graham B, Lee B, Shinawi M, Shen J, Kang SH, Pursley A, Lotze T, Kennedy G, Lansky-Shafer S, Weaver C, Roeder ER, Grebe TA, Arnold GL, Hutchison T, Reimschisel T, Amato S, Geragthy MT, Innis JW, Obersztyn E, Nowakowska B, Rosengren SS, Bader PI, Grange DK, Naqvi S, Garnica AD, Bernes SM, Fong CT, Summers A, Walters WD, Lupski JR, Stankiewicz P, Cheung SW and Patel A. Recurrent reciprocal 1q21. 1 deletions and duplications associated with microcephaly or macrocephaly and developmental and behavioral abnormalities. Nat Genet. **2008** Dec; 40(12):1466–1471. doi: 10.1038/ng.279. PMID:19029900.

Budnik V and Salinas PC. Wntsignalingduring synaptic development and plasticity. Curr Opin Neurobiol. **2011** Feb; 21(1):151–159. doi: 10.1016/j.conb. 2010.12. 002. Epub2011Jan 14. Review. PMID:21239163.

Buecker C and Wysocka J. Enhancers as information integration hubs in development: lessons from genomics. Trends Genet. **2012** Jun; 28(6):276–284. doi: 10.1016/j.tig.2012.02.008. Epub 2012 Apr 7.

Butler-Tjaden NE and Trainor PA. The developmental etiology and pathogenesis of Hirschsprung disease. Transl Res. **2013** Jul; 162(1):1–15. doi: 10.1016/j. trsl. 2013. 03. 001. Epub 2013 Mar 22. PMID: 23528997.

Butterfield RJ, Stevenson TJ, Xing L, Newcomb TM, Nelson B, Zeng W, Li X, Lu HM, Lu H, Farwell Gonzalez KD, Wei JP, Chao EC, Prior TW, Snyder PJ, Bonkowsky JL and Swoboda KJ. Congenital lethal motor neuron disease with a novel defect in ribosome biogenesis. Neurology. **2014** Apr 15; 82(15):1322–1330. doi: 10.1212/WNL.0000000000000305. Epub 2014 Mar 19. PMID:24647030.

Byers PH and Pyott SM. Recessively inherited forms of osteogenesis imperfecta. Annu Rev Genet. **2012**; 46:97. doi: 10.1146/annurev-genet-110711-155608. Review. PMID: 23145505.

Byrne JA, Simonsson S, Western PS and Gurdon JB. Nuclei of adult mammalian somatic cells are directly reprogrammed to oct-4 stem cell gene expression by amphibian oocytes. Curr Biol. **2003** Jul 15; 13(14):1206–1213. PMID: 12867031.

C

Caspersson T, Zech L, Johansson C and Modest EJ. Identification of human chromosomes by DNA-binding fluorescent agents. Chromosoma. **1970**; 30(2):215–227. PMID: 4193398.

Castle CD, Cassimere EK and Denicourt C. LAS1L interacts with the mammalian Rix1 complex to regulate ribosome biogenesis. Mol Biol Cell. **2012** Feb; 23(4):716–728. doi: 10.1091/mbc.E11-06-0530. Epub 2011 Dec 21. PMID:22190735.

Chambers I and Smith A. Self-renewal of teratocarcinoma and embryonic stem cells. Oncogene. **2004** Sep 20; 23(43):7150–7160. Review. PMID: 15378075.

Chan DC. Fusionand fission: interlinked processes critical for mitochondrial health. Annu Rev Genet. **2012**; 46:265–287. doi: 10.1146/annurev-genet-110410-132529. Epub2012Aug 29. Review. PMID:22934639.

Chassot AA, Gregoire EP, Magliano M, Lavery R and Chaboissier MC. Genetics of ovarian differentiation: Rspo1, a major player. Sex Dev. **2008**; 2(4–5): 219–227. doi: 10.1159/000152038. Epub 2008 Nov 5. PMID: 18987496.

Chen F, Pruett-Miller SM, Huang Y, Gjoka M, Duda K, Taunton J, Collingwood TN, Frodin M and Davis GD. High-frequency genome editing using ssDNA oligonucleotides with zinc-finger nucleases. Nat Methods. **2011** Jul 17; 8(9):753–755. doi: 10.1038/nmeth.1653. PMID: 21765410.

Chen M and Manley JL. Mechanisms of alternative splicing regulation: insights from molecular and genomics approaches. Nat Rev Mol Cell Biol. **2009** Nov; 10(11):741–754. doi: 10.1038/nrm2777. Epub 2009 Sep 23. Review. PMID: 19773805.

Cheng MC, Lu CL, Luu SU, Tsai HM, Hsu SH, Chen TT and Chen CH. Genetic and functional analysis of the DLG4 gene encoding the post-synaptic density protein 95 in schizophrenia. PLoS One. **2010** Dec 2; 5(12):e15107. doi: 10.1371/journal.pone. 0015107. PMID:21151988.

Cherry AB and Daley GQ. Reprogrammed cells for disease modeling and regenerative medicine. Annu Rev Med. **2013**; 64:277–290. doi: 10.1146/annurev-med-050311-163324. Review. PMID: 23327523.

Christensen ST, Clement CA, Satir P and Pedersen LB. Primary cilia and coordination of receptor tyrosine kinase (RTK) signalling. J Pathol. **2012** Jan; 226(2):172–184. doi: 10.1002/path.3004. Epub 2011 Nov 21. Review. PMID: 21956154.

Clausen TD, Mathiesen ER, Hansen T, Pedersen O, Jensen DM, Lauenborg J, Damm P. High prevalence of type 2 diabetes and pre-diabetes in adult offspring of women with gestational diabetes mellitus or type 1 diabetes: the role of intrauterine hyperglycemia. Diabetes Care. **2008** Feb; 31(2):340–346. Epub 2007 Nov 13. PMID:18000174.

Clayton PT. Disorders of bile acid synthesis. J Inherit Metab Dis. **2011** Jun; 34(3):593–604. doi: 10.1007/s10545-010-9259-3. Epub 2011 Jan 13. Review. PMID: 21229319.

Clynes D, Higgs DR and Gibbons RJ. The chromatin remodeller ATRX: a repeat offender in human disease. Trends Biochem Sci. **2013** Sep; 38(9): 461–466. doi: 10.1016/j.tibs.2013.06.011. Epub 2013 Aug 1. PMID: 23916100.

Comeaux MS, Wang J, Wang G, Kleppe S, Zhang VW, Schmitt ES, Craigen WJ, Renaud D, Sun Q and Wong LJ. Biochemical, molecular, and clinical diagnoses of patients with cerebral creatine deficiency syndromes. Mol Genet Metab. **2013** Jul; 109(3):260–268. doi: 10.1016/j.ymgme.2013.04.006. Epub 2013 Apr 17. PMID: 23660394.

Conaway RC and Conaway JW. The Mediator complex and transcription elongation. Biochim Biophys Acta. **2013** Jan; 1829(1):69–75. doi: 10.1016/j.bbagrm.2012.08.017. Epub 2012 Sep 13. Review. PMID: 22983086.

CoppAJ, Stanier P and Greene ND. Neuraltubedefects: recent advances, unsolved questions, and controversies. Lancet Neurol. **2013** Aug; 12(8):799–810.doi: 10.1016/S1474-4422(13)70110-8. Epub2013Jun 19. PMID:23790957.

Corrigan N, Brazil DP and McAuliffe F. Fetal cardiac effects of maternal hyperglycemia during pregnancy. Birth Defects Res A Clin Mol Teratol. **2009** Jun; 85(6):523–530. doi: 10.1002/bdra.20567. PMID:19180650.

Cottereau E, Mortemousque I, Moizard MP, Bürglen L, Lacombe D, Gilbert-Dussardier B, Sigaudy S, Boute O, David A, Faivre L, Amiel J, Robertson R, Viana Ramos F, Bieth E, Odent S, Demeer B, Mathieu M, Gaillard D, Van Maldergem L, Baujat G, Maystadt I, Héron D, Verloes A, Philip N, Cormier-Daire V, Frouté MF, Pinson L, Blanchet P, Sarda P, Willems M, Jacquinet A, Ratbi I, Van Den Ende J, Lackmy-Port Lis M, Goldenberg A, Bonneau D, Rossignol S and Toutain A. Phenotypic spectrum of Simpson–Golabi–Behmel syndrome in a series of 42 cases with a mutation in GPC3 and review of the literature. Am J Med Genet C Semin Med Genet. **2013** May; 163C(2):92–105. doi: 10.1002/ajmg.c.31360. Epub 2013 Apr 18. Review. PMID: 23606591.

Coufal NG, Garcia-Perez JL, Peng GE, Yeo GW, Mu Y, Lovci MT, Morell M, O'Shea KS, Moran JV and Gage FH. L1 retrotransposition in human neural progenitor cells. Nature. **2009** Aug 27; 460(7259):1127–1131. doi: 10.1038/nature08248. Epub 2009 Aug 5. PMID: 19657334.

Covey MV, Streb JW, Spektor R, Ballas N and Hammer RE. REST regulates the pool size of the different neural lineages by restricting the generation of neurons and oligodendrocytes from neural stem/progenitor cells. Development. **2012** Aug; 139(16):2878–2890. doi: 10.1242/dev.074765. Epub 2012 Jul 12. PMID: 22791895.

Croce JC and McClay DR. Evolution of the Wnt pathways. Methods Mol Biol. **2008**; 469:3–18. doi: 10.1007/978-1-60327-469-2. PMID:19109698.

D

Dannenberg LO, Chen HJ and Edenberg HJ. GATA-2 and HNF-3beta regulate the human alcohol dehydrogenase 1A (ADH1A) gene. DNA Cell Biol. **2005** Sep; 24(9):543–552. PMID: 16153155.

Deierlein AL, Siega-Riz AM, Chantala K, Herring AH. The association between maternal glucose concentration and child BMI at age 3 years. Diabetes Care. **2011**Feb; 34(2):480–4. doi: 10.2337/dc10-1766. Epub2011Jan 7. PMID: 21216858.

de Laat W and Duboule D. Topology of mammalian developmental enhancers and their regulatory landscapes. Nature. **2013** Oct 24; 502(7472):499–506. doi: 10.1038/nature12753. Review. PMID: 24153303.

Derti A, Garrett-Engele P, Macisaac KD, Stevens RC, Sriram S, Chen R, Rohl CA, Johnson JM and Babak T. A quantitative atlas of polyadenylation in five mammals. Genome Res. **2012** Jun; 22(6):1173–1183. doi: 10.1101/gr.132563.111. Epub 2012 Mar 27. PMID: 22454233.

Dietz HC. Marfan syndrome. In: Pagon RA, Adam MP, Ardinger HH, Bird TD, Dolan CR, Fong CT, Smith RJH and Stephens K (eds.), *GeneReviews®*

[Internet]. Seattle (WA): University of Washington, Seattle; 1993–2014. 2001 Apr 18 [updated 2014 Jun 12]. PMID: 20301510.

Dixon MJ, Marazita ML, Beaty TH and Murray JC. Cleft lip and palate: underst anding genetic and environmental influences. Nat Rev Genet. 2011 Mar; 12(3):167–178. doi: 10.1038/nrg2933. Review. PMID: 21331089.

Dixon PH, Trongwongsa P, Abu-Hayyah S, Ng SH, Akbar SA, Khawaja NP, Seck MJ, Savage PM and Fisher RA. Mutations in NLRP7 are associated with diploid biparental hydatidiform moles, but not and rogenetic complete moles. J Med Genet. **2012** Mar; 49(3):206–211. doi: 10.1136/jmedgenet-2011-100602. Epub 2012 Feb 7. PMID: 22315435.

Docherty LE, Kabwama S, Lehmann A, Hawke E, Harrison L, Flanagan SE, Ellard S, Hattersley AT, Shield JP, Ennis S, Mackay DJ and Temple IK. Clinical presentation of 6q24 transient neonatal diabetes mellitus (6q24 TNDM) and genotype–phenotype correlation in an international cohort of patients. Diabetologia. **2013** Apr; 56(4):758–762. doi: 10.1007/s00125-013-2832-1. Epub 2013 Feb 6. PMID: 23385738.

Doege CA, Inoue K, Yamashita T, Rhee DB, Travis S, Fujita R, Guarnieri P, Bhagat G, Vanti WB, Shih A, Levine RL, Nik S, Chen EI and Abeliovich A. Early-stage epigenetic modification during somatic cell reprogramming by Parp1 and Tet2. Nature. **2012** Aug 30; 488(7413):652–655. doi:10.1038/nature11333. PMID: 22902501.

Donald KA, Samia P, Kakooza-Mwesige A and Bearden D. Pediatric cerebral palsy in Africa: a systematic review. Semin Pediatr Neurol. **2014** Mar; 21(1):30–35. doi: 10.1016/j.spen.2014.01.001. Epub 2014 Jan 8. PMID: 24655402.

Doulatov S, Notta F, Laurenti E and Dick JE. Hematopoiesis: a human perspective. Cell Stem Cell. **2012** Feb 3; 10(2):120–136. doi: 10.1016/j.stem.2012.01.006. Review. PMID: 22305562.

Duester G. Retinoic acid synthesis and signaling during early organogenesis. Cell. **2008** Sep 19; 134(6):921–31. doi: 10.1016/j.cell.2008.09.002. Review. PMID: 18805086.

Dunlop EA and Tee AR. The kinase triad, AMPK, mTORC1 and ULK1, maintains energy and nutrient homoeostasis. Biochem Soc Trans. **2013** Aug; 41(4): 939–943. doi: 10.1042/BST20130030. Review. PMID: 23863160.

Dwivedi PP, Lam N and Powell BC. Boning up on glypicans — opportunities for new insights into bone biology. Cell Biochem Funct. **2013** Mar; 31(2): 91–114. doi: 10.1002/cbf.2939. Epub 2013 Jan 7. Review. PMID: 23297043.

Dyment DA, Sawyer SL, Chardon JW and Boycott KM. Recent advances in the genetic etiology of brain malformations. Curr Neurol Neurosci Rep. **2013** Aug; 13(8): 364. doi: 10.1007/s11910-013-0364-1. PMID: 23793931.

E

Eggermann T, Algar E, Lapunzina P, Mackay D, Maher ER, Mannens M, Netchine I, Prawitt D, Riccio A, Temple IK, Weksberg R. Clinical utility gene card for: Beckwith-Wiedemann Syndrome. Eur J Hum Genet. **2014** Mar; 22(3). doi: 10.1038/ejhg.2013.132. Epub 2013 Jul 3. PMID: 23820480.

Elkon R, Ugalde AP and Agami R. Alternative cleavage and polyadenylation: extent, regulation and function. Nat Rev Genet. **2013** Jul; 14(7):496–506. doi: 10.1038/nrg3482. Review. PMID: 23774734.

Erlinger S, Arias IM and Dhumeaux D. Inherited disorders of bilirubin transport and conjugation: new insights into molecular mechanisms and consequences. Gastroenterology. **2014** Jun; 146(7):1625–1638. doi: 10.1053/j.gastro.2014.03.047. Epub 2014 Apr 1. PMID: 24704527.

Ernst A, Alkass K, Bernard S, Salehpour M, Perl S, Tisdale J, Possnert G, Druid H and Frisén J. Neurogenesis in the striatum of the adult human brain. Cell. **2014** Feb 27; 156(5):1072–1083. doi: 10.1016/j.cell.2014.01.044. Epub 2014 Feb 20.

Evans MJ and Kaufman MH. Establishment in culture of pluripotential cells from mouse embryos. Nature. **1981** Jul 9; 292(5819):154–156. PMID: 7242681.

F

Fahed AC, Gelb BD, Seidman JG and Seidman CE. Genetics of congenital heart disease: the glass half empty. Circ Res. **2013** Feb 15; 112(4):707–720. Doi: 10.1161/CIRCRESAHA.112.300853. PMID: 23410880.

Faigle R and Song H. Signaling mechanisms regulating adult neural stem cells and neurogenesis. Biochim Biophys Acta. **2013** Feb; 1830(2):2435–2448. doi: 10.1016/j. bbagen. 2012. 09. 002. Epub 2012 Sep 12. PMID: 22982587.

Faulkner GJ, Kimura Y, Daub CO, Wani S, Plessy C, Irvine KM, Schroder K, Cloonan N, Steptoe AL, Lassmann T, Waki K, Hornig N, Arakawa T, Takahashi H, Kawai J, Forrest AR, Suzuki H, Hayashizaki Y, Hume DA, Orlando V, Grimmond SM and Carninci P. The regulated retrotransposon transcriptome of mammalian cells. Nat Genet. **2009** May; 41(5):563–571. doi: 10.1038/ng.368. Epub 2009 Apr 19. PMID: 19377475.

Fedor MJ. Alternative splicing minireview series: combinatorial control facilitates splicing regulation of gene expression and enhances genome diversity. J Biol Chem. **2008** Jan 18; 283(3):1209–1210. Epub 2007 Nov 16. PMID: 18024424.

Fernández LA, Sanz-Rodriguez F, Blanco FJ, Bernabéu C and Botella LM. Hereditary hemorrhagic telangiectasia, a vascular dysplasia affecting the

TGF-beta signaling pathway. Clin Med Res. **2006** Mar; 4(1):66–78. Review. PMID: 16595794.

Fernandez LA, Sanz-Rodriguez F, Zarrabeitia R, Perez-Molino A, Morales C, Restrepo CM, Ramirez JR, Coto E, Lenato GM, Bernabeu C and Botella LM. Mutation study of Spanish patients with hereditary hemorrhagic telangiectasia and expression analysis of endoglin and ALK1. Hum Mutat. **2006** Mar; 27(3):295. PMID: 16470589.

Ferrero GB, Baldassarre G, Panza E, Valenzise M, Pippucci T, Mussa A, Pepe E, Seri M and Silengo MC. A heritable cause of cleft lip and palate — Van der Woude syndrome caused by a novel IRF6 mutation. Review of the literature and of the differential diagnosis. Eur J Pediatr. **2010** Feb; 169(2): 223–228. doi: 10.1007/s00431-009-1011-3. Epub 2009 Jun 18. PMID: 19536562.

Ficicioglu C and Bearden D. Isolated neonatal seizures: when to suspect inborn errors of metabolism. Pediatr Neurol. **2011** Nov; 45(5):283–291. doi 10.1016/j. pediatrneurol. 2011. 07. 006. PMID: 22000307.

Fields RD. Neuroscience. Change in the brain's white matter. Science. **2010** Nov 5; 330(6005):768–769. doi: 10.1126/science. 1199139. PMID: 21051624.

Fields RD. Neuroscience: Map the other brain. Nature. **2013** Sep 5; 501(7465): 25–27. doi:10.1038/501025a. PMID:24010144.

Fisher RA, Khatoon R, Paradinas FJ, Roberts AP and Newlands ES. Repetitive complete hydatidiform mole can be biparental in origin and either male or female. Hum Reprod. **2000** Mar; 15(3):594–598. PMID: 10686202.

Fisher JP and Tweddle DA. Neonatal neuroblastoma. Semin Fetal Neonatal Med. **2012** Aug; 17(4):207–215. doi: 10.1016/j. siny. 2012. 05. 002. Epub 2012 Jun 4. PMID: 22673527.

Fitzpatrick E, Mitry RR and Dhawan A. Human hepatocyte transplantation: state of the art. J Intern Med. **2009** Oct; 266(4):339–357. doi: 10.1111/j. 1365–2796. 2009. 02152. x. Review. PMID: 19765179.

Fogel BL, Wexler E, Wahnich A, Friedrich T, Vijayendran C, Gao F, Parikshak N, Konopka G, Geschwind DH. RBFOX1 regulates both splicing and transcriptional networks in human neuronal development. Hum Mol Genet. **2012** Oct 1; 21(19): 4171–4186. doi: 10.1093/hmg/dds240. Epub 2012 Jun 23. PMID:22730494.

Frantz C, Stewart KM and Weaver VM. The extracellular matrix at a glance. J Cell Sci. **2010** Dec 15; 123(Pt 24):4195–4200. doi: 10.1242/jcs. 023820. PMID: 21123617.

Froese DS and Gravel RA. Genetic disorders of vitamin B_{12} metabolism: eight complementation groups — eight genes. Expert Rev Mol Med. **2010** Nov 29; 12:e37. doi: 10.1017/S1462399410001651. Review. PMID: 21114891.

Fulton D, Paez PM and Campagnoni AT. The multiple roles of myelin protein genes during the development of the oligodendrocyte. ASN Neuro. **2010** Feb 1; 2(1):e00027. doi: 10.1042/AN20090051. PMID:20017732.

G

Gao Z, Ure K, Ding P, Nashaat M, Yuan L, Ma J, Hammer RE and Hsieh J. The master negative regulator REST/NRSF controls adult neurogenesis by restraining the neurogenic program in quiescent stem cells. J Neurosci. **2011** Jun 29; 31(26):9772–9786. doi: 10.1523/JNEUROSCI. 1604-11. 2011.

Gaspard N, Bouschet T, Hourez R, Dimidschstein J, Naeije G, van den Ameele J, Espuny-Camacho I, Herpoel A, Passante L, Schiffmann SN, Gaillard A and Vanderhaeghen P. An intrinsic mechanism of corticogenesis from embryonicstem cells. Nature. **2008** Sep 18; 455(7211):351–357. doi: 10.1038/nature07287. Epub 2008 Aug 17. PMID:18716623.

Gauthier BR, Brun T, Sarret EJ, Ishihara H, Schaad O, Descombes P and Wollheim CB. Oligonucleotide microarray analysis reveals PDX1 as an essential regulator of mitochondrial metabolism in rat islets. J Biol Chem. **2004** Jul 23; 279(30):31121–31130. Epub 2004 May 19.

Gibson EM, Purger D, Mount CW, Goldstein AK, Lin GL, Wood LS, Inema I, Miller SE, Bieri G, Zuchero JB, Barres BA, Woo PJ, Vogel H and Monje M. Neuronal activity promotes oligodendrogenesis and adaptive myelination in the mammalian brain. Science. **2014** May 2; 344(6183):1252304. doi: 10.1126/science.1252304. Epub 2014 Apr 10. PMID: 24727982.

Gifford CA, Ziller MJ, Gu H, Trapnell C, Donaghey J, Tsankov A, Shalek AK, Kelley DR, Shishkin AA, Issner R, Zhang X, Coyne M, Fostel JL, Holmes L, Meldrim J, Guttman M, Epstein C, Park H, Kohlbacher O, Rinn J, Gnirke A, Lander ES, Bernstein BE and Meissner A. Transcriptional and epigenetic dynamics during specification of human embryonic stem cells. Cell. **2013** May 23; 153(5):1149–1163. doi: 10.1016/j.cell.2013.04.037. Epub 2013 May 9. PMID: 23664763.

Gilmore EC and Walsh CA. Genetic causes of microcephaly and lessons for neuronal development. Wiley Interdiscip Rev Dev Biol. **2013** Jul; 2(4):461–478. doi: 10.1002/wdev.89. Epub 2012 Oct 4. PMID: 24014418.

Gimelli S, Caridi G, Beri S, McCracken K, Bocciardi R, Zordan P, Dagnino M, Fiorio P, Murer L, Benetti E, Zuffardi O, Giorda R, Wells JM, Gimelli G and Ghiggeri GM. Mutations in SOX17 are associated with congenital anomalies of the kidney and the urinary tract. Hum Mutat. **2010** Dec; 31(12): 1352–1359. doi: 10.1002/humu.21378. Epub 2010 Nov 9. PMID: 20960469.

Goumy C, Gouas L, Marceau G, Coste K, Veronese L, Gallot D, Sapin V, Vago P and Tchirkov A. Retinoid pathway and congenital diaphragmatic hernia: hypothesis from the analysis of chromosomal abnormalities. Fetal Diagn Ther. **2010**; 28(3):129–139. doi: 10.1159/000313331. Epub 2010 May 26. Review. PMID: 20501978.

Gower WA and Nogee LM. Surfactant dysfunction. Paediatr Respir Rev. **2011** Dec; 12(4):223–229. doi: 10.1016/j.prrv.2011.01.005. Epub 2011 Mar 5. PMID: 22018035.

Grayson BE, Seeley RJ and Sandoval DA. Wired on sugar: the role of the CNS in the regulation of glucose homeostasis. Nat Rev Neurosci. **2013** Jan; 14(1): 24–37. doi: 10.1038/nrn3409. Epub 2012 Dec 12. Review. PMID:23232606.

Green AJ, Smith M and Yates JR. Loss of heterozygosity on chromosome 16p13. 3 in hamartomas from tuberous sclerosis patients. Nat Genet. **1994** Feb; 6(2):193–196. PMID: 8162074.

Greer PL, Hanayama R, Bloodgood BL, Mardinly AR, Lipton DM, Flavell SW, Kim TK, Griffith EC, Waldon Z, Maehr R, Ploegh HL, Chowdhury S, Worley PF, Steen J and Greenberg ME. The Angelman syndrome protein Ube3A regulates synapse development by ubiquitinating arc. Cell. **2010** Mar 5; 140(5):704–716. doi: 10.1016/j.cell.2010.01.026. PMID:20211139.

Greer EL and Shi Y. Histone methylation: a dynamic mark in health, disease and inheritance. Nat Rev Genet. **2012** Apr 3; 13(5):343–357. doi: 10.1038/nrg 3173. PMID: 22473383.

Gregg, NM. Congenital cataract following German measles in the mother. 1941. Aust N Z J Opthalmol. **1991** Nov; 19(4): 267–276. PMID: 1789963.

Greig LC, Woodworth MB, Galazo MJ, Padmanabhan H and Macklis JD. Molecular logic of neocortical projection neuron specification, development and diversity. Nat Rev Neurosci. **2013** Nov; 14(11):755–769. doi: 10.1038/ nrn3586. Epub 2013 Oct 9. PMID: 4105342.

Grosen D, Chevrier C, Skytthe A, Bille C, Mølsted K, Sivertsen A, Murray JC and Christensen K. A cohort study of recurrence patterns among more than 54,000 relatives of oral cleft cases in Denmark: support for the multifactorial threshold model of inheritance. J Med Genet. **2010a** Mar; 47(3):162–168. doi: 10.1136/jmg.2009.069385. Epub 2009 Sep 14. PMID: 1975216.

Grosen D, Bille C, Pedersen JK, Skytthe A, Murray JC and Christensen K. Recurrence risk for offspring of twins discordant for oral cleft: a population-based cohort study of the Danish 1936–2004 cleft twin cohort. Am J Med Genet A. **2010b** Oct; 152A(10):2468–2474. doi: 10.1002/ajmg.a.33608. PMID: 20799319.

Guntur AR and Rosen CJ. The skeleton: a multi-functional complex organ: new insights into osteoblasts and their role in bone formation: the central role of PI3 Kinase. J Endocrinol. **2011** Nov; 211(2):123–130. doi: 10.1530/JOE-11-0175. Epub 2011 Jun 14. PMID: 21673026.

Guo C, Sun Y, Guo C, MacDonald BT, Borer JG and Li X. Dkk1 in the pericloaca mesenchyme regulates formation of anorectal and genitourinary tracts. Dev Biol. **2014** Jan 1; 385(1):41–51. PMID: 24479159.

H

Hafner C, Toll A, Gantner S, Mauerer A, Lurkin I, Acquadro F, Fernández-Casado A, Zwarthoff EC, Dietmaier W, Baselga E, Parera E, Vicente A, Casanova A, Cigudosa J, Mentzel T, Pujol RM, Landthaler M and Real FX. Keratinocytic epidermal nevi are associated with mosaic RAS mutations. J Med Genet. **2012** Apr; 49(4):249–253. doi: 10.1136/jmedgenet-2011-100637. PMID: 22499344.

Hafner C and Groesser L. Mosaic rasopathies. Cell Cycle. **2013** Jan 1; 12(1): 43–50. doi: 10.4161/cc.23108. Epub 2012 Dec 19. Review. PMID: 23255105.

Hagood JS and Ambalavanan N. Systems biology of lung development and regeneration: current knowledge and recommendations for future research. Wiley Interdiscip Rev Syst Biol Med. **2013** Mar–Apr; 5(2):125–133. doi: 10.1002/wsbm.1205. Epub 2013 Jan 4. Review. PMID: 23293056.

Hallman M. The surfactant system protects both fetus and newborn. Neonatology. **2013**; 103(4):320–326. doi: 10.1159/000349994. Epub 2013 May 31. Review. PMID: 23736009.

Han H, Irimia M, Ross PJ, Sung HK, Alipanahi B, David L, Golipour A, Gabut M, Michael IP, Nachman EN, Wang E, Trcka D, Thompson T, O'Hanlon D, Slobodeniuc V, Barbosa-Morais NL, Burge CB, Moffat J, Frey BJ, Nagy A, Ellis J, Wrana JL and Blencowe BJ. MBNL proteins repress ES-cell-specific alternative splicing and reprogramming. Nature. **2013** Jun 13; 498(7453): 241–245. doi: 10.1038/nature12270. Epub 2013 Jun 5. PMID: 23739326.

Hannan NR, Fordham RP, Syed YA, Moignard V, Berry A, Bautista R, Hanley NA, Jensen KB and Vallier L. Generation of multipotent foregut stem cells from human pluripotent stem cells. Stem Cell Reports. **2013** Oct 10; 1(4): 293–306. doi: 10.1016/j.stemcr.2013.09.003. eCollection 2013. PMID: 24319665.

Hanson D, Stevens A, Murray PG, Black GC and Clayton PE. Identifying biological pathways that underlie primordial short stature using network analysis. J Mol Endocrinol. **2014** Jun; 52(3):333–344. doi: 10.1530/JME-14-0029. Epub 2014 Apr 7. PMID: 24711643.

Helin K and Dhanak D. Chromatin proteins and modifications as drug targets. Nature. **2013** Oct 24; 502(7472):480–488. doi: 10.1038/nature12751.

Heller M and Burd L. Review of ethanol dispersion, distribution, and elimination from the fetal compartment. Birth Defects Res A Clin Mol Teratol. **2014** Apr; 100(4): 277–283. doi: 10.1002/bdra.23232. Epub 2014 Mar 10. PMID: 24616297.

Herriges M and Morrisey EE. Lung development: orchestrating the generation and regeneration of a complex organ. Development. **2014** Feb; 141(3): 502–513. doi: 10.1242/dev.098186. PMID: 24449833.

Higuchi R, Sugimoto T, Tamura A, Kioka N, Tsuno Y, Higa A, Yoshikawa N. Early features in neuroimaging of two siblings with molybdenum cofactor deficiency. Pediatrics. **2014** Jan; 133(1):e267–271. doi: 10.1542/peds.2013-0935. Epub 2013 Dec 30. PMID:24379235.

Hildebrandt F, Benzing T and Katsanis N. Ciliopathies. N Engl J Med. **2011** Apr 21; 364(16):1533–1543. doi: 10.1056/NEJMra1010172. Review. PMID: 21506742.

Howson CP, Christianson AC and Modell B. Controlling birth defects: reducing the hidden toll of dying and disabled children in lower-income countries. **2008**. Washington (District of Columbia): Disease Control Priorities Project.

Hu J, Srivastava K, Wieland M, Runge A, Mogler C, Besemfelder E, Terhardt D, Vogel MJ, Cao L, Korn C, Bartels S, Thomas M and Augustin HG. Endothelial cell-derived angiopoietin-2 controls liver regeneration as a spatiotemporal rheostat. Science. **2014** Jan 24; 343(6169):416–419. doi: 10.1126/science.1244880. PMID: 24458641.

Huxley A. Uncle Spencer, In Little Mexican. **1924**: 64.

Hyman BT and Yuan J. Apoptotic and non-apoptotic roles of caspases in neuronal physiology and pathophysiology. Nat Rev Neurosci. **2012** May 18; 13(6): 395–406. doi: 10.1038/nrn3228. PMID: 22595785.

Hynes RO and Naba A. Overview of the matrisome — an inventory of extracellular matrix constituents and functions. Cold Spring Harb Perspect Biol. **2012** Jan 1; 4(1):a004903. doi: 10.1101/cshperspect.a004903. Review. PMID: 21937732.

I

Iasevoli F, Tomasetti C and de Bartolomeis A. Scaffolding proteins of the post-synaptic density contribute to synaptic plasticity by regulating receptor localization and distribution: relevance for neuropsychiatric diseases. Neurochem Res. **2013** Jan; (1):1–22. doi: 10.1007/s11064-012-0886-y. Epub 2012 Sep 19. Review. PMID: 22991141.

Iliff JJ, Wang M, Liao Y, Plogg BA, Peng W, Gundersen GA, Benveniste H, Vates GE, Deane R, Goldman SA, Nagelhus EA and Nedergaard M. A paravascular pathway facilitates CSF flow through the brain parenchyma and the clearance of interstitial solutes, including amyloid β. Sci Transl Med. **2012** Aug 15; 4(147):147ra111. doi: 10.1126/scitranslmed. 3003748. PMID: 22896675.

Imbard A, Benoist JF and Blom HJ. Neural tube defects, folic acid and methylation. Int J Environ Res Public Health. **2013** Sep 17; 10(9):4352–4389. doi: 10.3390/ijerph10094352.

Inoue S, Cho BH, Song CH, Fujimiya M, Murakami G and Matsubara A. Migration and distribution of neural crest-derived cells in the human adrenal cortex at 9–16 weeks of gestation: an immunohistochemical study. Okajimas Folia Anat Jpn. **2010** May; 87(1):11–16. PMID: 20715567.

Ishimoto H and Jaffe RB. Development and function of the human fetal adrenal cortex: a key component in the feto-placental unit. Endocr Rev. **2011** Jun; 32(3):317–355. doi: 10.1210/er. 2010-0001. Epub 2010 Nov 4. PMID: 21051591.

Itoh N and Ornitz DM. Fibroblast growth factors: from molecular evolution to roles in development, metabolism and disease. J Biochem. **2011** Feb; 149(2):121–130. doi: 10.1093/jb/mvq121. Epub 2010 Oct 12. PMID: 20940169.

J

Janoueix-Lerosey I, Lequin D, Brugières L, Ribeiro A, de Pontual L, Combaret V, Raynal V, Puisieux A, Schleiermacher G, Pierron G, Valteau-Couanet D, Frebourg T, Michon J, Lyonnet S, Amiel J and Delattre O. Somatic and germline activating mutations of the ALK kinase receptor in neuroblastoma. Nature. **2008** Oct 16; 455(7215):967–970. doi: 10.1038/nature07398. PMID: 18923523.

Jeste SS and Geschwind DH. Disentangling the heterogeneity of autism spectrum disorder through genetic findings. Nat Rev Neurol. **2014** Feb; 10(2):74–81. doi: 10.1038/nrneurol.2013.278. Epub 2014 Jan 28. PMID: 24468882.

Jiang Y, Tsai TF, Bressler J and Beaudet AL. Imprinting in Angelman and Prader–Willi syndromes. Curr Opin Genet Dev. **1998** Jun; 8(3):334–342. PMID: 9691003.

Jones PA. Functions of DNA methylation: islands, start sites, gene bodies and beyond. Nat Rev Genet. **2012** May 29; 13(7):484–492. doi: 10.1038/nrg3230. PMID: 22641018.

K

Kamil D, Tepelmann J, Berg C, Heep A, Axt-Fliedner R, Gembruch U and Geipel A. Spectrum and outcome of prenatally diagnosed fetal tumors. Ultrasound Obstet Gynecol. **2008** Mar; 31(3):296–302. doi: 10.1002/uog.5260. PMID: 1830720.

Kanamori T, Kanai MI, Dairyo Y, Yasunaga K, Morikawa RK and Emoto K. Compartmentalized calcium transients trigger dendrite pruning in Drosophila sensory neurons. Science. **2013** Jun 21; 340(6139):1475–1478. doi: 10.1126/science.1234879. Epub 2013 May 30. PMID: 23722427.

Karbstein K. Quality control mechanisms during ribosome maturation. Trends Cell Biol. **2013** May; 23(5):242–250. doi: 10.1016/j.tcb.2013.01.004. Epub 2013 Feb 1. Review. PMID: 23375955.

Kawaji H, Frith MC, Katayama S, Sandelin A, Kai C, Kawai J, Carninci P and Hayashizaki Y. Dynamic usage of transcription start sites within core promoters. Genome Biol. **2006**; 7(12):R118. PMID: 17156492.

Kawamata R, Suzuki Y, Yada Y, Koike Y, Kono Y, Yada T and Takahashi N. Gut hormone profiles in preterm and term infants during the first 2 months of life. J Pediatr Endocrinol Metab. **2014** Jul 1; 27(7–8):717–723. doi: 10.1515/jpem-2013-0385. PMID: 24572982.

Keitges EA, Pasion R, Burnside RD, Mason C, Gonzalez-Ruiz A, Dunn T, Masiello M, Gebbia JA, Fernandez CO and Risheg H. Prenatal diagnosis of two fetuses with deletions of 8p23.1, critical region for congenital diaphragmatic hernia and heart defects. Am J Med Genet A. **2013** Jul; 161A(7):1755–1758. doi: 10.1002/ajmg.a.35965. Epub 2013 May 21. PMID: 23696316.

Kelberman D and Dattani MT. Hypothalamic and pituitary development: novel insights into the aetiology. Eur J Endocrinol. **2007** Aug; 157(Suppl 1): S3–S14. Review. PMID: 17785694.

Kelsom C and Lu W. Development and specification of GABAergic cortical interneurons. Cell Biosci. **2013** Apr 23; 3(1):19. doi: 10.1186/2045-3701-3-19. PMID: 23618463.

Kempná P and Flück CE. Adrenal gland development and defects. Best Pract Res Clin Endocrinol Metab. **2008** Feb; 22(1):77–93. doi: 10.1016/j.beem.2007. 07.008. PMID: 18279781.

Kennedy M, Awong G, Sturgeon CM, Ditadi A, LaMotte-Mohs R, Zúñiga-Pflücker JC and Keller G. T lymphocyte potential marks the emergence of definitive hematopoietic progenitors in human pluripotent stem cell differentiation cultures. Cell Rep. **2012** Dec 27; 2(6):1722–1735. doi: 10.1016/j.celrep. 2012.11.003. Epub 2012 Dec 7. PMID: 21544903.

Kerac M, Postels DG, Mallewa M, Alusine Jalloh A, Voskuijl WP, Groce N, Gladstone M and Molyneux E. The interaction of malnutrition and neurologic disability in Africa. Semin Pediatr Neurol. **2014** Mar; 21(1):42–49. doi: 10.1016/j.spen.2014.01.003. Epub 2014 Jan 4. PMID: 24655404.

Kerzendorfer C, Colnaghi R, Abramowicz I, Carpenter G and O'Driscoll M. Meier–Gorlin syndrome and Wolf–Hirschhorn syndrome: two developmental disorders highlighting the importance of efficient DNA replication for normal development and neurogenesis. DNA Repair (Amst). **2013** Aug; 12(8):637–644. doi: 10.1016/j.dnarep.2013.04.016. Epub 2013 May 23. PMID: 23706772.

Keys C, Drewett M and Burge DM. Gastroschisis: the cost of an epidemic. J Pediatr Surg. **2008** Apr; 43(4):654–657. doi:10.1016/j.jpedsurg.2007.12.005. PMID: 18405711.

Khalid O, Kim JJ, Kim HS, Hoang M, Tu TG, Elie O, Lee C, Vu C, Horvath S, Spigelman I and Kim Y. Gene expression signatures affected by alcohol-induced DNA methylomic deregulation in human embryonic stem cells. Stem Cell Res. **2014** May; 12(3):791–806. doi: 10.1016/j.scr.2014.03.009. Epub 2014 Apr 12. PMID: 24751885.

Khoo C, Yang J, Weinrott SA, Kaestner KH, Naji A, Schug J and Stoffers DA. Research resource: the pdx1 cistrome of pancreatic islets. Mol Endocrinol. **2012** Mar; 26(3):521–533. doi: 10.1210/me.2011-1231. Epub 2012 Feb 9. PMID: 22322596.

Kim MS, Pinto SM, Getnet D, Nirujogi RS, Manda SS, Chaerkady R, Madugundu AK, Kelkar DS, Isserlin R, Jain S, Thomas JK, Muthusamy B, Leal-Rojas P, Kumar P, Sahasrabuddhe NA, Balakrishnan L, Advani J, George B, Renuse S, Selvan LD, Patil AH, Nanjappa V, Radhakrishnan A, Prasad S, Subbannayya T, Raju R, Kumar M, Sreenivasamurthy SK, Marimuthu A, Sathe GJ, Chavan S, Datta KK, Subbannayya Y, Sahu A, Yelamanchi SD, Jayaram S, Rajagopalan P, Sharma J, Murthy KR, Syed N, Goel R, Khan AA, Ahmad S, Dey G, Mudgal K, Chatterjee A, Huang TC, Zhong J, Wu X, Shaw PG, Freed D, Zahari MS, Mukherjee KK, Shankar S, Mahadevan A, Lam H, Mitchell CJ, Shankar SK, Satishchandra P, Schroeder JT, Sirdeshmukh R, Maitra A, Leach SD, Drake CG, Halushka MK, Prasad TS, Hruban RH, Kerr CL, Bader GD, Iacobuzio-Donahue CA, Gowda H and Pandey A. Nature. **2014** May 29; 509(7502):575–581. doi: 10.1038/nature13302. PMID: 24870542.

Kleinsmith LJ and Pierce GB Jr. Multipotentiality of single embryonal carcinoma cells Cancer Res. **1964** Oct; 24:1544–1551. PMID: 14234000.

Knapp KM, Brogly SB, Muenz DG, Spiegel HM, Conway DH, Scott GB, Talbot JT, Shapiro DE and Read JS. Prevalence of congenital anomalies in infants with in utero exposure to antiretrovirals. Pediatr Infect Dis J. **2012** Feb; 31(2):164–170. doi: 10.1097/INF.0b013e318235c7aa.

Knöll R, Postel R, Wang J, Krätzner R, Hennecke G, Vacaru AM, Vakeel P, Schubert C, Murthy K, Rana BK, Kube D, Knöll G, Schäfer K, Hayashi T, Holm T, Kimura A, Schork N, Toliat MR, Nürnberg P, Schultheiss HP, Schaper W, Schaper J, Bos E, Den Hertog J, van Eeden FJ, Peters PJ, Hasenfuss G, Chien KR and Bakkers J. Laminin-alpha4 and integrin-linked kinase mutations cause human cardiomyopathy *via* simultaneous defects in cardiomyocytes and endothelial cells. Circulation. **2007** Jul 31; 116(5): 515–525. Epub 2007 Jul 23. PMID: 17646580.

Koga M, Ishiguro H, Yazaki S, Horiuchi Y, Arai M, Niizato K, Iritani S, Itokawa M, Inada T, Iwata N, Ozaki N, Ujike H, Kunugi H, Sasaki T, Takahashi M, Watanabe Y, Someya T, Kakita A, Takahashi H, Nawa H, Muchardt C, Yaniv M and Arinami T. Involvement of SMARCA2/BRM in the SWI/SNF chro-matin-remodeling complex in schizophrenia. Hum Mol Genet. **2009** Jul 1; 18(13):2483–2494. doi: 10.1093/hmg/ddp166. Epub 2009 Apr 10. PMID: 19363039.

Kohli RM and Zhang Y. TET enzymes, TDG and the dynamics of DNA demeth-ylation. Nature. **2013** Oct 24; 502(7472):472–479. doi: 10.1038/nature12750.

Komada M. Sonic hedgehog signaling coordinates the proliferation and differen-tiation of neural stem/progenitor cells by regulating cell cycle kinetics during development of the neocortex Congenit Anom (Kyoto). **2012** Jun; 52(2):72–77. doi: 10.1111/j.1741-4520.2012.00368.x. PMID:22639991.

Kopp P. Mutations in the Pendred syndrome (PDS/SLC26A) gene: an increas-ingly complex phenotypic spectrum from goiter to thyroid hypoplasia. J Clin Endocrinol Metab. **2014** Jan; 99(1):67–69. doi: 10.1210/jc.2013-4319. PMID: 24384016.

Kouzarides T. Chromatin modifications and their function. Cell. **2007** Feb 23; 128(4):693–705. Review. PMID: 17320507.

Krakow D and Rimoin DL. The skeletal dysplasias. Genet Med. **2010** Jun; 12(6):327–341. doi: 10.1097/GIM.0b013e3181daae9b. Review. PMID: 20556869.

Krone N and Arlt W. Genetics of congenital adrenal hyperplasia. Best Pract Res Clin Endocrinol Metab. **2009** Apr; 23(2):181–192. doi: 10.1016/j.beem.2008. 10.014. Review. PMID: 19500762.

L

Laine CM, Joeng KS, Campeau PM, Kiviranta R, Tarkkonen K, Grover M, Lu JT, Pekkinen M, Wessman M, Heino TJ, Nieminen-Pihala V, Aronen M, Laine T, Kröger H, Cole WG, Lehesjoki AE, Nevarez L, Krakow D, Curry CJ, Cohn DH, Gibbs RA, Lee BH and Mäkitie O. WNT1 mutations in early-onset osteoporosis and osteogenesis imperfecta. N Engl J Med. **2013** May 9; 368(19):1809–1816. doi: 10.1056/NEJMoa1215458. PMID: 23656646.

Lake JI and Heuckeroth RO. Enteric nervous system development: migration, differentiation, and disease. Am J Physiol Gastrointest Liver Physiol. **2013** Jul 1; 305(1):G1–G24. doi: 10.1152/ajpgi.00452.2012. Epub 2013 May 2. Review. PMID: 236398.

Lancaster MA, Renner M, Martin CA, Wenzel D, Bicknell LS, Hurles ME, Homfray T, Penninger JM, Jackson AP and Knoblich JA. Cerebralorganoids model human brain development and microcephaly. Nature. **2013** Sep 19; 501(7467):373–379. doi: 10.1038/nature12517. Epub 2013 Aug 28. PMID: 23995685.

Larsen. *Textbook of Embryology*. Schoenwolf GC, Bleyl SB, Prauer PR, Francis WP (eds.). **2009**; Published Elsevier 2008. Paperback 2009.

Lattanzi W, Bukvic N, Barba M, Tamburrini G, Bernardini C, Michetti F and Di Rocco C. Genetic basis of single-suture synostoses: genes, chromosomes and clinical implications. Childs Nerv Syst. **2012** Sep; 28(9):1301–1310. doi: 10.1007/s00381-012-1781-1. Epub 2012 Aug 8. Review. PMID: 22872241.

Laurenti E, Doulatov S, Zandi S, Plumb I, Chen J, April C, Fan JB and Dick JE. The transcriptional architecture of early human hematopoiesis identifies multilevel control of lymphoid commitment. Nat Immunol. **2013** Jul; 14(7):756–763. doi: 10.1038/ni.2615. Epub 2013 May 26. PMID: 23708252.

Leder P. Discontinuous genes. N Engl J Med. **1978** May 11; 298(19):1079–1081. PMID: 643015.

Lee JS, Ward WO, Knapp G, Ren H, Vallanat B, Abbott B, Ho K, Karp SJ and Corton JC. Transcriptional ontogeny of the developing liver. BMC Genomics. **2012** Jan 19; 13:33. doi: 10.1186/1471-2164-13-33. PMID: 22260730.

Lewin B. *Genes VII*. **2000**; Oxford:Oxford University Press. **1999**

Lewis DA, Curley AA, Glausier JR and Volk DW. Cortical parvalbumin interneurons and cognitive dysfunction in schizophrenia. Trends Neurosci. **2012** Jan; 35(1):57–67. doi: 10.1016/j.tins.2011.10.004. Epub 2011 Dec 6. Review. PMID: 22154068.

Licatalosi DD and Darnell RB. RNA processing and its regulation: global insights into biological networks. Nat Rev Genet. **2010** Jan; 11(1):75–87. doi: 10.1038/nrg2673. PMID: 20019688.

Lister R, Mukamel EA, Nery JR, Urich M, Puddifoot CA, Johnson ND, Lucero J, Huang Y, Dwork AJ, Schultz MD, Yu M, Tonti-Filippini J, Heyn H, Hu S, Wu JC, Rao A, Esteller M, He C, Haghighi FG, Sejnowski TJ, Behrens MM and Ecker JR. Global epigenomic reconfiguration during mammalian brain development. Science. **2013** Aug 9; 341(6146):1237905. doi: 10.1126/science.1237905. Epub 2013 Jul 4. PMID: 23828890.

Liu JS, Schubert CR, Fu X, Fourniol FJ, Jaiswal JK, Houdusse A, Stultz CM, Moores CA and Walsh CA. Molecular basis for specific regulation of neuronal kinesin-3 motors by doublecortin family proteins. Mol Cell. **2012** Sep 14; 47(5):707–721. doi: 10.1016/j.molcel.2012.06.025. Epub 2012 Aug 1. PMID: 22857951.

Liu XA, Rizzo V and Puthanveettil SV. Pathologies of axonal transport in neurodegenerative diseases. Transl Neurosci. **2012** Dec 1; 3(4):355–372. PMID: 23750323.

Loh KM and Lim B. Epigenetics: actors in the cell reprogramming drama. Comment on early-stage epigenetic modification during somatic cell reprogramming by Parp1 and Tet2. Nature. **2012** Aug 30; 488(7413): 599–600. doi: 10.1038/488599a. PMID: 22932382.

Lossi L and Merighi A. *In vivo* cellular and molecular mechanisms of neuronal apoptosis in the mammalian CNS. Prog Neurobiol. **2003** Apr; 69(5):287–312. PMID: 12787572.

Lu J and Clark AG. Impact of microRNA regulation on variation in human gene expression. Genome Res. **2012** Jul; 22(7):1243–1254. doi: 10.1101/gr.132514.111. Epub 2012 Mar 28. PMID: 22456605.

Lu R and Wang GG. Tudor: a versatile family of histone methylation 'readers'. Trends Biochem Sci. **2013** Nov; 38(11):546–555. doi: 10.1016/j.tibs.2013.08.002. Epub 2013 Sep 10. Review. PMID: 24035451.

Lu W, Zhang Y, Liu D, Songyang Z and Wan M. Telomeres-structure, function, and regulation. Exp Cell Res. **2013** Jan 15; 319(2):133–141. doi:10.1016/j.yexcr.2012.09.005. Epub 2012 Sep 21. PMID: 23006819.

Lubinsky M. Hypothesis: estrogen related thrombosis explains the pathogenesis and epidemiology of gastroschisis. Am J Med Genet A. **2012** Apr; 158A(4): 808–811. doi: 10.1002/ajmg.a.35203. Epub 2012 Mar 1. Review. PMID: 22383174.

Lubinsky M. A vascular and thrombotic model of gastroschisis. Am J Med Genet A. **2014** Apr; 164A(4):915–917. doi: 10.1002/ajmg.a.36370. Epub 2014 Jan 23. PMID: 24458365.

Luco RF and Misteli T. More than a splicing code: integrating the role of RNA, chromatin and non-coding RNA in alternative splicing regulation. Curr Opin

Genet Dev. **2011** Aug; 21(4):366–372. doi: 10.1016/j.gde.2011.03.004. Epub 2011 Apr 15. Review. PMID: 21497503.

Luscan A, Laurendeau I, Malan V, Francannet C, Odent S, Giuliano F, Lacombe D, Touraine R, Vidaud M, Pasmant E and Cormier-Daire V. Mutations in SETD2 cause a novel overgrowth condition. J Med Genet. **2014** May 22; 102402. doi: 10.1136/jmedgenet-2014-102402. [Epub ahead of print]. PMID: 24852293.

M

Ma DK, Marchetto MC, Guo JU, Ming GL, Gage FH and Song H. Epigenetic choreographers of neurogenesis in the adult mammalian brain. Nat Neurosci. **2010** Nov; 13(11):1338–1344. doi: 10.1038/nn.2672. PMID: 20975758.

Maass PG, Rump A, Schulz H, Stricker S, Schulze L, Platzer K, Aydin A, Tinschert S, Goldring MB, Luft FC and Bähring S. A misplaced lncRNA causes brachydactyly in humans. J Clin Invest. **2012** Nov 1; 122(11): 3990–4002. doi: 10.1172/JCI65508. Epub 2012 Oct 24. PMID: 23093776.

Mackie EJ, Tatarczuch L and Mirams M. The skeleton: a multi-functional complex organ: the growth plate chondrocyte and endochondral ossification. J Endocrinol. **2011** Nov; 211(2):109–121. doi: 10.1530/JOE-11-0048. Epub 2011 Jun 3. Review. PMID: 21642379.

Magistri M, Faghihi MA, St Laurent G 3rd and Wahlestedt C. Regulation of chromatin structure by long noncoding RNAs: focus on natural antisense transcripts. Trends Genet. **2012** Aug; 28(8):389–396. doi: 10.1016/j.tig. 2012.03.013. Epub 2012 Apr 26. Review. PMID: 22541732.

Mahadevan S, Wen S, Wan YW, Peng HH, Otta S, Liu Z, Iacovino M, Mahen EM, Kyba M, Sadikovic B and Van den Veyver IB. NLRP7 affects trophoblast lineage differentiation, binds to overexpressed YY1 and alters CpG methylation. Hum Mol Genet. **2014** Feb 1; 23(3):706–716. doi: 10.1093/hmg/ ddt457. Epub 2013 Sep 18. PMID: 24105472.

Mangelsdorf DJ, Thummel C, Beato M, Herrlich P, Schütz G, Umesono K, Blumberg B, Kastner P, Mark M, Chambon P and Evans RM. The nuclear receptor superfamily: the second decade. Cell. **1995** Dec 15; 83(6):835–839. PMID: 8521507.

Mangold E, Ludwig KU and Nöthen MM. Breakthroughs in the genetics of orofacial clefting. Trends Mol Med. **2011** Dec; 17(12):725–733. doi:10.1016/j.molmed.2011.07.007. Epub 2011 Aug 30. Review. PMID: 21885341.

Marín O. Interneuron dysfunction in psychiatric disorders. **Nat Rev Neurosci. 2012** Jan 18; 13(2):107–120. doi: 10.1038/nrn3155. PMID: 22251963.

Maris JM. Recent advances in neuroblastoma. N Engl J Med. **2010** Jun 10; 362(23):2202–2211. doi: 10.1056/NEJMra0804577. Review. PMID: 2055837.

Martin GR. Isolation of a pluripotent cell line from early mouse embryos cultured in medium conditioned by teratocarcinoma stem cells. Proc Natl Acad Sci USA. **1981** Dec; 78(12):7634–7638. PMID: 6950406.

Mason I. Initiation to end point: the multiple roles of fibroblast growth factors in neural development. Nat Rev Neurosci. **2007** Aug; 8(8):583–596. Review. PMID: 17637802.

Massagué J and Xi Q. TGF-β control of stem cell differentiation genes. FEBS Lett. **2012** Jul 4; 586(14):1953–1958. doi: 10.1016/j.febslet.2012.03.023. Epub 2012 Apr 10. Review. PMID: 22710171.

Mattick JS. RNA driving the epigenetic bus. EMBO J. **2012** Feb 1; 31(3):515–516. doi: 10.1038/emboj.2011.479. Epub 2012 Feb 1. PMID: 22293829.

McGrath J and Solter D. Completion of mouse embryogenesis requires both the maternal and paternal genomes. Cell. **1984** May; 37(1):179–183. PMID: 6722870.

Mefford HC, Shafer N, Antonacci F, Tsai JM, Park SS, Hing AV, Rieder MJ, Smyth MD, Speltz ML, Eichler EE and Cunningham ML. Copy number variation analysis in single-suture craniosynostosis: multiple rare variants including RUNX2 duplication in two cousins with metopic craniosynostosis. Am J Med Genet A. **2010** Sep; 152A(9):2203–2210. doi: 10.1002/ajmg.a. 33557. PMID: 20683987.

Meier JJ, Köhler CU, Alkhatib B, Sergi C, Junker T, Klein HH, Schmidt WE and Fritsch H. Beta-cell development and turnover during prenatal life in humans. Eur J Endocrinol. **2010** Mar; 162(3):559–568. doi: 10.1530/EJE-09-1053. Epub 2009 Dec 18. PMID: 20022941.

Mellis DJ, Itzstein C, Helfrich MH and Crockett JC. The skeleton: a multi-functional complex organ: the role of key signalling pathways in osteoclast differentiation and in bone resorption. J Endocrinol. **2011** Nov; 211(2): 131–143. doi: 10.1530/JOE-11-0212. Epub 2011 Sep 8. PMID: 21903860.

Mikkola HK and Orkin SH. The journey of developing hematopoietic stem cells. Development. **2006** Oct; 133(19):3733–3744. PMID: 16968814.

Miller JA, Ding SL, Sunkin SM, Smith KA, Ng L, Szafer A, Ebbert A, Riley ZL, Royall JJ, Aiona K, Arnold JM, Bennet C, Bertagnolli D, Brouner K, Butler S, Caldejon S, Carey A, Cuhaciyan C, Dalley RA, Dee N, Dolbeare TA, Facer BA, Feng D, Fliss TP, Gee G, Goldy J, Gourley L, Gregor BW, Gu G, Howard RE, Jochim JM, Kuan CL, Lau C, Lee CK, Lee F, Lemon TA, Lesnar P, McMurray B, Mastan N, Mosqueda N, Naluai-Cecchini T, Ngo NK, Nyhus J, Oldre A, Olson E, Parente J, Parker PD, Parry SE, Stevens A, Pletikos M,

Reding M, Roll K, Sandman D, Sarreal M, Shapouri S, Shapovalova NV, Shen EH, Sjoquist N, Slaughterbeck CR, Smith M, Sodt AJ, Williams D, Zöllei L, Fischl B, Gerstein MB, Geschwind DH, Glass IA, Hawrylycz MJ, Hevner RF, Huang H, Jones AR, Knowles JA, Levitt P, Phillips JW, Sestan N, Wohnoutka P, Dang C, Bernard A, Hohmann JG and Lein ES. Transcriptional landscape of the prenatal human brain. Nature. **2014** Apr 10; 508(7495):199–206. doi: 10.1038/nature13185. Epub 2014 Apr 2. PMID: 24695229.

Missler M, Südhof TC and Biederer T. Synaptic cell adhesion. Cold Spring Harb Perspect Biol. **2012** Apr 1; 4(4):a005694. doi: 10.1101/cshperspect.a005694. PMID: 2227866.

Mizushima N and Levine B. Autophagy in mammalian development and differentiation. Nat Cell Biol. **2010** Sep; 12(9):823–830. doi: 10.1038/ncb0910-823. Review. PMID: 20811354.

Modarresi F, Faghihi MA, Lopez-Toledano MA, Fatemi RP, Magistri M, Brothers SP, van der Brug MP and Wahlestedt C. Inhibition of natural antisense transcripts in vivo results in gene-specific transcriptional upregulation. Nat Biotechnol. **2012** Mar 25; 30(5):453–459. doi: 10.1038/nbt.2158. PMID: 22446693.

Momeni P, Glöckner G, Schmidt O, von Holtum D, Albrecht B, Gillessen-Kaesbach G, Hennekam R, Meinecke P, Zabel B, Rosenthal A, Horsthemke B and Lüdecke HJ. Mutations in a new gene, encoding a zinc-finger protein, cause tricho-rhino-phalangeal syndrome type I. Nat Genet. **2000** Jan; 24(1): 71–74. PMID: 10615131.

Monod J. Genetic regulatory mechanisms in the synthesis of proteins. J Mol Biol. **1961** Jun; 3:318–356. PMID: 13718526.

Moretti A, Bellin M, Welling A, Jung CB, Lam JT, Bott-Flügel L, Dorn T, Goedel A, Höhnke C, Hofmann F, Seyfarth M, Sinnecker D, Schömig A and Laugwitz KL. Patient-specific induced pluripotent stem-cell models for long-QT syndrome. N Engl J Med. **2010** Oct 7; 363(15):1397–1409. doi: 10.1056/NEJMoa0908679. Epub 2010 Jul 21. PMID: 20660394.

Morey L and Helin K. Polycomb group protein-mediated repression of transcription. Trends Biochem Sci. **2010** Jun; 35(6):323–332. doi: 10.1016/j.tibs. 2010.02.009. Epub 2010 Mar 24. PMID: 20346678.

Mossé YP, Laudenslager M, Longo L, Cole KA, Wood A, Attiyeh EF, Laquaglia MJ, Sennett R, Lynch JE, Perri P, Laureys G, Speleman F, Kim C, Hou C, Hakonarson H, Torkamani A, Schork NJ, Brodeur GM, Tonini GP, Rappaport E, Devoto M and Maris JM. Identification of ALK as a major familial neuroblastoma predisposition gene. Nature. **2008** Oct 16; 455(7215):930–935. doi: 10.1038/nature07261. Epub 2008 Aug 24. PMID: 18724359.

Mou H, Zhao R, Sherwood R, Ahfeldt T, Lapey A, Wain J, Sicilian L, Izvolsky K, Musunuru K, Cowan C and Rajagopal J. Generation of multipotent lung and airway progenitors from mouse ESCs and patient-specific cystic fibrosis iPSCs. Cell Stem Cell. **2012** Apr 6; 10(4):385–397. doi: 10.1016/j.stem. 2012.01.018. Erratum in: Cell Stem Cell. 2012 May 4; 10(5):635. PMID: 22482504.

Moustakas A and Heldin CH. Coordination of TGF-β signaling by ubiquitylation. Mol Cell. **2013** Sep 12; 51(5):555–556. doi: 10.1016/j.molcel.2013.08.034. PMID: 24034692.

Muotri AR, Chu VT, Marchetto MC, Deng W, Moran JV and Gage FH. Somatic mosaicism in neuronal precursor cells mediated by L1 retrotransposition. Nature. **2005** Jun 16; 435(7044):903–910. PMID: 15959507.

Mullis PE. Genetics of isolated growth hormone deficiency. J Clin Res Pediatr Endocrinol. **2010**; 2(2):52–62. doi: 10.4274/jcrpe.v2i2.52. Epub 2010 May 1. Review. PMID: 21274339.

Mummery CL, Zhang J, Ng ES, Elliott DA, Elefanty AG and Kamp TJ. Differentiation of human embryonic stem cells and induced pluripotent stem cells to cardiomyocytes: a methods overview. Circ Res. **2012** Jul 20; 111(3):344–358. doi: 10.1161/CIRCRESAHA.110.227512. PMID: 22821908.

Musselman CA, Lalonde ME, Côté J and Kutateladze TG. Perceiving the epigenetic landscape through histone readers. Nat Struct Mol Biol. **2012** Dec; 19(12):1218–1227. doi: 10.1038/nsmb. 2436. Review. PMID: 23211769.

N

Narendra D, Walker JE and Youle R. Mitochondrial quality control mediated by PINK1 and Parkin: links to parkinsonism. Cold Spring Harb Perspect Biol. **2012** Nov 1; 4(11): a011338. doi: 10.1101/cshperspect.a011338. PMID: 23125018.

Narisawa A, Komatsuzaki S, Kikuchi A, Niihori T, Aoki Y, Fujiwara K, Tanemura M, Hata A, Suzuki Y, Relton CL, Grinham J, Leung KY, Partridge D, Robinson A, Stone V, Gustavsson P, Stanier P, Copp AJ, Greene ND, Tominaga T, Matsubara Y and Kure S. Mutations in genes encoding theglycinecleavage system predispose toneural tubedefects in mice and humans. Hum Mol Genet. **2012** Apr 1; 21(7):1496–1503. doi: 10.1093/hmg/ddr585. Epub 2011 Dec 13.

Nelson SB, Sugino K and Hempel CM. The problem of neuronal cell types: a physiological genomics approach. Trends Neurosci. **2006** Jun; 29(6):339–345. Epub 2006 May 22. Review. PMID: 16714064.

Netchine I, Rossignol S, Azzi S, Brioude F and Le Bouc Y. Imprinted anomalies in fetal and childhood growth disorders: the model of Russell-Silver and Beckwith-Wiedemann syndromes. Endocr Dev. **2012**; 23:60–70. doi: 10.1159/000341750. Epub 2012 Nov 23. PMID: 23182821.

New MI. An update of congenital adrenal hyperplasia. Ann N Y Acad Sci. **2004** Dec; 1038:14–43. PMID: 15838095.

Nichol PF, Corliss RF, Yamada S, Shiota K and Saijoh Y. Muscle patterning in mouse and human abdominal wall development and omphalocele specimens of humans. Anat Rec (Hoboken). **2012** Dec; 295(12):2129–2140. doi: 10.1002/ar.22556. Epub 2012 Sep 14. PMID: 22976993.

Nikopoulos K, Venselaar H, Collin RW, Riveiro-Alvarez R, Boonstra FN, Hooymans JM, Mukhopadhyay A, Shears D, van Bers M, de Wijs IJ, van Essen AJ, Sijmons RH, Tilanus MA, van Nouhuys CE, Ayuso C, Hoefsloot LH and Cremers FP. Overview of the mutation spectrum in familial exudative vitreoretinopathy and Norrie disease with identification of 21 novel variants in FZD4, LRP5, and NDP. Hum Mutat. **2010** Jun; 31(6):656–666. doi: 10.1002/humu.21250. PMID: 20340138.

Nogueira C, Almeida LS, Nesti C, Pezzini I, Videira A, Vilarinho L and Santorelli FM. Syndromes associated with mitochondrial DNA depletion. Ital J Pediatr. **2014** Apr 3; 40:34. doi: 10.1186/1824-7288-40-34. PMID: 24708634.

Nyhan WL, Barshop BA, Ozand PT. Atlas of Metabolic Diseases 2nd edition. **2005** Hodder Arnold, Great Britain.

O

Obermeier B, Daneman R and Ransohoff RM. Development, maintenance and disruption of the blood-brain barrier. Nat Med. **2013** Dec; 19(12):1584–1596. doi: 10.1038/nm. 3407. Epub 2013 Dec 5. PMID: 24309662.

O'Driscoll M, Dobyns WB, van Hagen JM and Jeggo PA. Cellular and clinical impact of haploinsufficiency for genes involved in ATR signaling. Am J Hum Genet. **2007** Jul; 81(1):77–86. Epub 2007 May 17. PMID: 17564965.

P

Paget J. Lancet. **1882**; 2: 1017.

Pagliarini DJ, Calvo SE, Chang B, Sheth SA, Vafai SB, Ong SE, Walford GA, Sugiana C, Boneh A, Chen WK, Hill DE, Vidal M, Evans JG, Thorburn DR, Carr SA and Mootha VK. A mitochondrial protein compendium elucidates complex I disease biology. Cell. **2008** Jul 11; 134(1):112–123. doi: 10.1016/j.cell.2008.06.016. PMID: 18614015.

Pagnamenta AT, Lise S, Harrison V, Stewart H, Jayawant S, Quaghebeur G, Deng AT, Murphy VE, Sadighi Akha E, Rimmer A, Mathieson I, Knight SJ, Kini U, Taylor JC and Keays DA. Exome sequencing can detect pathogenic mosaic mutations present at low allele frequencies. J Hum Genet. **2012** Jan; 57(1):70–72. doi: 10.1038/jhg.2011.128. Epub 2011 Dec 1. PMID: 22129557.

Pal S, Gupta R, Kim H, Wickramasinghe P, Baubet V, Showe LC, Dahmane N and Davuluri RV. Alternative transcription exceeds alternative splicing in generating the transcriptome diversity of cerebellar development. Genome Res. **2011** Aug; 21(8):1260–1272. doi: 10.1101/gr. 120535.111. Epub 2011 Jun 28. PMID: 21712398.

Pallade GE. **The Nobel Prize Lecture of George E. Palade**. Nobel lecture Intracellular aspects of the process of protein secretion. 1974; The Nobel Foundation: **ISBN 981-02-0791-3**.

Palmieri F. Mitochondrial transporters of the SLC25 family and associated diseases: a review. J Inherit Metab Dis. **2014** Jul; 37(4):565–575. doi: 10.1007/s10545-014-9708-5. Epub **2014** May 6. PMID: 24797559.

Pera MF and Tam PP. Extrinsic regulation of pluripotent stem cells. Nature. **2010** Jun 10; 465(7299):713–720. doi: 10.1038/nature09228. PMID: 20535200.

Parry S, Zhang H, Biggio J, Bukowski R, Varner M, Xu Y, Andrews WW, Saade GR, Esplin MS, Leite R, Ilekis J, Reddy UM, Sadovsky Y, Blair IA. Maternal serum serpin B7 is associated with early spontaneous preterm birth. Am J Obstet Gynecol. **2014** Jun 19. pii: S0002-9378(14)00608-5. doi: 10.1016/j. ajog.2014.06.035. [Epub ahead of print] PMID:24954659.

Pineda-Alvarez DE, Roessler E, Hu P, Srivastava K, Solomon BD, Siple CE, Fan CM and Muenke M. Missense substitutions in the GAS1 protein present in holoprosencephaly patients reduce the affinity for its ligand, SHH. Hum Genet. **2012** Feb; 131(2):301–310. doi: 10.1007/s00439-011-1078-6. Epub 2011 Aug 13. PMID: 21842183.

Pingault V, Ente D, Dastot-Le Moal F, Goossens M, Marlin S and Bondurand N. Review and update of mutations causing Waardenburg syndrome. Hum Mutat. **2010** Apr; 31(4):391–406. doi: 10.1002/humu.21211. PMID: 20127975.

Pirinen E, Lo Sasso G and Auwerx J. Mitochondrial sirtuins and metabolic homeostasis. Best Pract Res Clin Endocrinol Metab. **2012** Dec; 26(6):759–770. doi: 10.1016/j.beem.2012.05.001. Epub 2012 May 31. PMID: 23168278.

Plagemann A. Maternal diabetes and perinatal programming. Early Hum Dev. **2011** Nov; 87(11):743–747. doi: 10.1016/j. earlhumdev. 2011. 08. 018. Epub 2011 Sep 23. PMID: 21945359.

Platt FM. Sphingolipid lysosomal storage disorders. Nature. **2014** Jun 5; 510(7503):68–75. doi: 10.1038/nature13476. PMID: 24899306.

Polizio AH, Chinchilla P, Chen X, Manning DR and Riobo NA. Sonic Hedgehog activates the GTPases Rac1 and RhoA in a Gli-independent manner through coupling of smoothened to Gi proteins. Sci Signal. **2011** Nov 22; 4(200):pt7. doi: 10.1126/scisignal. 2002396. Review. PMID: 22114142.

Porter JA, Young KE and Beachy PA. Cholesterol modification of hedgehog signaling proteins in animal development. Science. **1996** Oct 11; 274(5285): 255–259. Erratum in: Science 1996 Dec 6; 274(5293):1597. PMID: 8824192.

R

Rauen KA. The rasopathies. Annu Rev Genomics Hum Genet. **2013**; 14:355–369. doi: 10.1146/annurev-genom-091212-153523. Epub 2013 Jul 15. Review. PMID: 23875798.

Riedl SJ and Salvesen GS. The apoptosome: signalling platform of cell death. Nat Rev Mol Cell Biol. **2007** May; 8(5):405–413. Epub 2007 Mar 21. Review. PMID: 17377525.

Rifkin DB and Todorovic V. Bone matrix to growth factors: location, location, location. J Cell Biol. **2010** Sep 20; 190(6):949–951. doi: 10.1083/jcb. 201008116. PMID: 20855500.

Risch N, Hoffmann TJ, Anderson M, Croen LA, Grether JK and Windham GC. Familial recurrence of autism spectrum disorder: evaluating genetic and environmental contributions. Am J Psychiatry. **2014** Jun 27. doi: 10.1176/ appi.ajp.2014.13101359. [Epub ahead of print]. PMID: 24969362.

Reiner O and Sapir T. LIS1 functions in normal development and disease. Curr Opin Neurobiol. **2013** Dec; 23(6):951–956. doi: 10.1016/j.conb.2013.08.001. Epub 2013 Aug 23. PMID: 23973156.

Rodríguez-Seguí S, Akerman I and Ferrer J. GATA believe it: new essential regulators of pancreas development. J Clin Invest. **2012** Oct 1; 122(10):3469–3471. doi: 10.1172/JCI65751. Epub 2012 Sep 24. PMID: 23006323.

Roessler E, Ward DE, Gaudenz K, Belloni E, Scherer SW, Donnai D, Siegel-Bartelt J, Tsui LC and Muenke M. Cytogenetic rearrangements involving the loss of the Sonic Hedgehog gene at 7q36 cause holoprosencephaly. Hum Genet. **1997** Aug; 100(2):172–181. PMID: 9254845.

Romani M, Micalizzi A and Valente EM. Joubert syndrome: congenital cerebellar ataxia with the molar tooth. Lancet Neurol. **2013** Sep; 12(9):894–905. doi: 10.1016/S1474-4422(13)70136-4. Epub 2013 Jul 17. Review. PMID: 23870701.

Ronan JL, Wu W and Crabtree GR. From neural development to cognition: unexpected roles for chromatin. Nat Rev Genet. **2013** May; 14(5):347–359. doi: 10.1038/nrg3413. Epub 2013 Apr 9. Review. Erratum in: Nat Rev Genet. 2013 Jun; 14(6):440. PMID: 23568486.

S

Saftig P and Klumperman J. Lysosome biogenesis and lysosomal membrane proteins: trafficking meets function. Nat Rev Mol Cell Biol. **2009** Sep; 10(9): 623–635. doi: 10.1038/nrm2745. Epub 2009 Aug 12. Review. PMID 19672277.

Saito T, Takeda N, Amiya E, Nakao T, Abe H, Semba H, Soma K, Koyama K, Hosoya Y, Imai Y, Isagawa T, Watanabe M, Manabe I, Komuro I, Nagai R and Maemura K. VEGF-A induces its negative regulator, soluble form of VEGFR-1, by modulating its alternative splicing. FEBS Lett. **2013** Jul 11; 587(14):2179–2185.

Sakata K, Woo NH, Martinowich K, Greene JS, Schloesser RJ, Shen L and Lu B. Critical role of promoter IV-driven BDNF transcription in GABAergic transmission and synaptic plasticity in the prefrontal cortex. Proc Natl Acad Sci USA. **2009** Apr 7; 106(14):5942–5947. doi: 10.1073/pnas. 0811431106. Epub 2009 Mar 17. PMID: 19293383.

Sander JD and Joung JK. CRISPR-Cas systems for editing, regulating and targeting genomes. Nat Biotechnol. **2014** Apr; 32(4):347-355. doi: 10.1038/nbt. 2842. Epub **2014** Mar 2. PMID: 24584096.

Saudubray JM, Nassogne MC, de Lonlay P and Touati G. Clinical approach to inherited metabolic disorders in neonates: an overview. Semin Neonatol. **2002** Feb; 7(1):3–15. Review. PMID: 12069534.

Savastano CP, El-Jaick KB, Costa-Lima MA, Abath CM, Bianca S, Cavalcanti DP, Félix TM, Scarano G, Llerena JC Jr, Vargas FR, Moreira MÂ, Seuánez HN, Castilla EE and Orioli IM. Molecular analysis of holoprosencephaly in South America. Genet Mol Biol. **2014** Mar; 37(Suppl 1):250–262. PMID: 24764759.

Scarpulla RC. Metabolic control of mitochondrial biogenesis through the PGC-1 family regulatory network. Biochim Biophys Acta. **2011** Jul; 1813(7):1269–1278. doi: 10.1016/j. bbamcr.2010.09.019. Epub 2010 Oct 13. Review. PMID: 20933024.

Scheidecker S, Etard C, Pierce NW, Geoffroy V, Schaefer E, Muller J, Chennen K, Flori E, Pelletier V, Poch O, Marion V, Stoetzel C, Strähle U, Nachury MV and Dollfus H. Exome sequencing of Bardet–Biedl syndrome patient identifies a null mutation in the BBSome subunit BBIP1 (BBS18). J Med Genet. **2014** Feb; 51(2):132–136. doi: 10.1136/jmedgenet-2013-101785. Epub 2013 Sep 11. PMID: 24026985.

Schuettengruber B, Martinez AM, Iovino N and Cavalli G. Trithorax group proteins: switching genes on and keeping them active. Nat Rev Mol Cell Biol. **2011** Nov 23; 12(12):799–814. doi: 10.1038/nrm3230. Review. PMID: 22108599.

Scott JD and Pawson T. Cell signaling in space and time: where proteins come together and when they're apart. Science. **2009** Nov 27; 326(5957):1220–1224. doi: 10.1126/science. 1175668. PMID: 19965465.

Seabright M. Human chromosome banding. Lancet. **1972** Apr 29; 1(7757):967. PMID: 4112138.

Sepp T, Yates JR and Green AJ. Loss of heterozygosity in tuberous sclerosis hamartomas. J Med Genet. **1996** Nov; 33(11):962–964.

Settembre C, Fraldi A, Medina DL and Ballabio A. Signals from the lysosome: a control centre for cellular clearance and energy metabolism. Nat Rev Mol Cell Biol. **2013** May; 14(5):283–296. doi: 10.1038/nrm3565. PMID: 23609508.

Shea AK and Steiner M. Cigarette smoking during pregnancy. Nicotine Tob Res. **2008** Feb; 10(2):267–78. doi: 10.1080/14622200701825908. PMID: 18236291.

Shiang R, Thompson LM, Zhu YZ, Church DM, Fielder TJ, Bocian M, Winokur ST and Wasmuth JJ. Mutations in the transmembrane domain of FGFR3 cause the most common genetic form of dwarfism, achondroplasia. Cell. **1994** Jul 29; 78(2):335–342. PMID: 7913883.

Shirley MD, Tang H, Gallione CJ, Baugher JD, Frelin LP, Cohen B, North PE, Marchuk DA, Comi AM and Pevsner J. Sturge–Weber syndrome and port-wine stains caused by somatic mutation in GNAQ. N Engl J Med. **2013** May 23; 368(21):1971–1979. doi: 10.1056/NEJMoa1213507. Epub 2013 May 8. PMID: 23656586.

Simons M and Mlodzik M. Planar cell polarity signaling: from fly development to human disease. Annu Rev Genet. **2008**; 42:517–540. doi: 10.1146/annurev.genet.42.110807. 091432. PMID: 18710302.

Simonsson S and Gurdon JB. Changing cell fate by nuclear reprogramming. Cell Cycle. **2005** Apr; 4(4):513–515. Epub 2005 May 1. PMID: 15753660.

Simões-Costa M, Tan-Cabugao J, Antoshechkin I, Sauka-Spengler T and Bronner ME. Transcriptome analysis reveals novel players in the cranial neural crest gene regulatory network. Genome Res. **2014** Feb; 24(2):281–290. doi: 10.1101/gr. 161182. 113. Epub 2014 Jan 3. PMID: 24389048.

Si-Tayeb K, Lemaigre FP and Duncan SA. Organogenesis and development of the liver. Dev Cell. **2010** Feb 16; 18(2):175–189. doi: 10.1016/j.devcel. 2010.01.011. Review. PMID: 20159590.

Slavotinek AM. The genetics of common disorders — Congenital diaphragmatic hernia. Eur J Med Genet. **2014** May 2; S1769-7212(14): 00097-4. doi: 10.1016/j.ejmg.2014.04.012. [Epub ahead of print]. PMID: 24793812.

Smith DW. Recognizable Patterns of Human Malformation. **1974**. W.B. Saunders Company.

Smith M, Hopkinson DA and Harris H. Studies on the subunit structure and molecular size of the human alcohol dehydrogenase isozymes determined by the different loci, ADH1, ADH2, and ADH3. Ann Hum Genet. **1973a** Apr; 36(4):401–414. PMID: 4748759.

Smith M, Hopkinson DA and Harris H. Studies on the properties of the human alcohol dehydrogenase isozymes determined by the different loci ADH1, ADH2, ADH3. Ann Hum Genet. **1973b** Jul; 37(1):49–67. PMID: 4796765.

Smith M, Hopkinson DA and Harris H. Alcohol dehydrogenase isozymes in adult human stomach and liver: evidence for activity of the ADH 3 locus. Ann Hum Genet. **1972** Mar; 35(3):243–253. PMID: 5072686.

Smith M, Hopkinson DA and Harris H. Developmental changes and polymorphism in human alcohol dehydrogenase. Ann Hum Genet. **1971** Feb; 34(3): 251–271. PMID: 5548434.

Solomon BD, Bear KA, Wyllie A, Keaton AA, Dubourg C, David V, Mercier S, Odent S, Hehr U, Paulussen A, Clegg NJ, Delgado MR, Bale SJ, Lacbawan F, Ardinger HH, Aylsworth AS, Bhengu NL, Braddock S, Brookhyser K, Burton B, Gaspar H, Grix A, Horovitz D, Kanetzke E, Kayserili H, Lev D, Nikkel SM, Norton M, Roberts R, Saal H, Schaefer GB, Schneider A, Smith EK, Sowry E, Spence MA, Shalev SA, Steiner CE, Thompson EM, Winder TL, Balog JZ, Hadley DW, Zhou N, Pineda-Alvarez DE, Roessler E and Muenke M. Genotypic and phenotypic analysis of 396 individuals with mutations in Sonic Hedgehog. J Med Genet. **2012** Jul; 49(7):473–479. doi: 10.1136/jmedgenet-2012-101008. PMID: 22791840.

Soshnikova N. Dynamics of Polycomb and Trithorax activities during development. Birth Defects Res A Clin Mol Teratol. **2011** Aug; 91(8):781–787. doi: 10.1002/bdra. 20774. Epub 2011 Feb 2. Review. PMID: 21290568.

Sousa SB, Abdul-Rahman OA, Bottani A, Cormier-Daire V, Fryer A, Gillessen-Kaesbach G, Horn D, Josifova D, Kuechler A, Lees M, MacDermot K, Magee A, Morice-Picard F, Rosser E, Sarkar A, Shannon N, Stolte-Dijkstra I, Verloes A, Wakeling E, Wilson L and Hennekam RC. Nicolaides–Baraitser syndrome: delineation of the phenotype. Am J Med Genet A. **2009** Aug; 149A(8):1628–1640. doi: 10.1002/ajmg.a.32956. PMID: 19606471.

Spalding KL, Bergmann O, Alkass K, Bernard S, Salehpour M, Huttner HB, Boström E, Westerlund I, Vial C, Buchholz BA, Possnert G, Mash DC, Druid H and Frisén J. Dynamics of hippocampal neurogenesis in adult humans. Cell. **2013** Jun 6; 153(6):1219–1227. doi: 10.1016/j.cell.2013.05.002. PMID: 23746839.

Spitz F and Furlong EE. Transcription factors: from enhancer binding to developmental control. Nat Rev Genet. **2012** Sep; 13(9):613–626. doi: 10.1038/nrg 3207. Epub 2012 Aug 7. PMID: 22868264.

Stockler S, Plecko B, Gospe SM Jr, Coulter-Mackie M, Connolly M, van Karnebeek C, Mercimek-Mahmutoglu S, Hartmann H, Scharer G, Struijs E, Tein I, Jakobs C, Clayton P and Van Hove JL. Pyridoxine dependent epilepsy and antiquitin deficiency: clinical and molecular characteristics and recommendations for diagnosis, treatment and follow-up. Mol Genet Metab. **2011** Sep–Oct; 104(1–2):48–60. doi: 10.1016/j.ymgme.2011.05.014. Epub 2011 May 24. PMID: 21704546.

Strickland S and Mahdavi V. The induction of differentiation in teratocarcinoma stem cells by retinoic acid. Cell. **1978** Oct; 15(2):393–403. PMID: 214238.

Su JS, Tsai TF, Chang HM, Chao KM, Su TS and Tsai SF. Distant HNF1 site as a master control for the human class I alcohol dehydrogenase gene expression. J Biol Chem. **2006** Jul 21; 281(29):19809–19821. Epub 2006 May 4. PMID: 16675441.

Szinnai G. Genetics of normal and abnormal thyroid development in humans. Best Pract Res Clin Endocrinol Metab. **2014** Mar; 28(2):133–150. doi: 10.1016/j.beem.2013.08.005. Epub 2013 Aug 20. PMID: 24629857.

T

Takahashi K and Yamanaka S. Induction of pluripotent stem cells from mouse embryonic and adult fibroblast cultures by defined factors. Cell. **2006** Aug 25; 126(4):663–676. Epub 2006 Aug 10. PMID: 16904174.

Takahashi Y, Sipp D and Enomoto H. Tissue interactions in neural crest cell development and disease. Science. **2013** Aug 23; 341(6148):860–863. doi: 10.1126/science.1230717. PMID: 23970693.

Tan H, Yi L, Rote NS, Hurd WW and Mesiano S. Progesterone receptor-A and -B have opposite effects on proinflammatory gene expression in human myometrial cells: implications for progesterone actions in human pregnancy and parturition. J Clin Endocrinol Metab. **2012** May; 97(5):E719–E730. doi: 10.1210/jc.2011-3251. Epub 2012 Mar 14. PMID: 22419721.

Tartaglia M and Gelb BD. Disorders of dysregulated signal traffic through the RAS–MAPK pathway: phenotypic spectrum and molecular mechanisms. Ann N Y Acad Sci. **2010** Dec; 1214:99–121. doi: 10.1111/j.1749-6632.2010.05790. x. Epub 2010 Oct 19. Review. PMID: 20958325.

Tatton-Brown K, Trevor RP, Cole M and Rahman N. *Sotos Syndrome Synonym: Cerebral Gigantism Gene Reviews.* http://www.ncbi.nlm.nih.gov/books/ NBK1479.

Tatton-Brown K, Seal S, Ruark E, Harmer J, Ramsay E, Del Vecchio Duarte S, Zachariou A, Hanks S, O'Brien E, Aksglaede L, Baralle D, Dabir T, Gener B, Goudie D, Homfray T, Kumar A, Pilz DT, Selicorni A, Temple IK, Van Maldergem L and Yachelevich N. Mutations in the DNA methyltransferase gene DNMT3A cause an overgrowth syndrome with intellectual disability. Nat Genet. **2014** Apr; 46(4):385–388. doi: 10.1038/ng.2917. Epub 2014 Mar 9. PMID: 24614070.

Tau GZ and Peterson BS. Normal development of brain circuits. Neuropsychopharmacology. **2010** Jan; 35(1):147–168. doi: 10.1038/npp. 2009.115.Review. PMID: 19794405.

Taylor PD, Samuelsson AM and Poston L. Maternal obesity and the developmental programming of hypertension: a role for leptin. Acta Physiol (Oxf). **2014** Mar; 210(3):508–523. doi: 10.1111/apha. 12223. PMID: 24433239.

Tenenbaum-Rakover Y, Mamanasiri S, Ris-Stalpers C, German A, Sack J, Allon-Shalev S, Pohlenz J and Refetoff S. Clinical and genetic characteristics of congenital hypothyroidism due to mutations in the thyroid peroxidase (TPO) gene in Israelis. Clin Endocrinol (Oxf). **2007** May; 66(5):695–702. Epub 2007 Mar 23. PMID: 17381485.

Teng T, Thomas G, Mercer CA. Growth control and ribosomopathies. Curr Opin Genet Dev. **2013** Feb; 23(1):63–71. doi: 10.1016/j.gde.2013.02.001. Epub2013Mar 13. Review. PMID: 23490481.

Teresa-Rodrigo ME, Eckhold J, Puisac B, Dalski A, Gil-Rodríguez MC, Braunholz D, Baquero C, Hernández-Marcos M, de Karam JC, Ciero M, Santos-Simarro F, Lapunzina P, Wierzba J, Casale CH, Ramos FJ, Gillessen-Kaesbach G, Kaiser FJ and Pié J. Functional characterization of NIPBL physiological splice variants and eight splicing mutations in patients with Cornelia de Lange syndrome. Int J Mol Sci. **2014** Jun 10; 15(6):10350–10364. doi: 10.3390/ijms150610350. PMID: 24918291.

Thomson E, Ferreira-Cerca S and Hurt E. Eukaryotic ribosome biogenesis at a glance. J Cell Sci. **2013** Nov 1; 126(Pt 21):4815–4821. doi: 10.1242/jcs. 111948. PMID: 24172536.

Thomson JA, Itskovitz-Eldor J, Shapiro SS, Waknitz MA, Swiergiel JJ, Marshall VS and Jones JM. Embryonic stem cell lines derived from human blastocysts. Science. **1998** Nov 6; 282(5391):1145–1147. PMID: 9804556.

Tian B and Manley JL. Alternative cleavage and polyadenylation: the long and short of it. Trends Biochem Sci. **2013** Jun; 38(6):312–320. doi: 10.1016/j. tibs. 2013.03.005. Epub 2013 Apr 27.

Ting JT, Peça J and Feng G. Functional consequences of mutations in postsynaptic scaffolding proteins and relevance to psychiatric disorders. Annu Rev Neurosci. **2012**; 35:49–71. doi: 10.1146/annurev-neuro-062111-150442. Epub 2012 Apr 20. Review. PMID: 22540979.

Tiscornia G, Monserrat N and Izpisua-Belmonte JC. Modelling long QT syndrome with iPS cells: be still, my beating heart. Circ Res. **2011** Mar 18; 108(6):648–649. doi: 10.1161/RES.0b013e318216f0db. PMID: 21415406.

Tiscornia G, Vivas EL and Izpisúa-Belmonte JC. Diseases in a dish: modeling human genetic disorders using induced pluripotent cells. Nat Med. **2011** Dec; 17(12):1570–1576. doi: 10.1038/nm.2504. Review. PMID: 22146428.

Toledo-Rodriguez M, Goodman P, Illic M, Wu C and Markram H. Neuropeptide and calcium-binding protein gene expression profiles predictneuronalanatomical type in the juvenile rat. J Physiol. **2005** Sep 1; 567(Pt 2):401–413. Epub 2005 Jun 9. PMID: 15946970.

Tomaselli S, Megiorni F, De Bernardo C, Felici A, Marrocco G, Maggiulli G, Grammatico B, Remotti D, Saccucci P, Valentini F, Mazzilli MC, Majore S and Grammatico P. Syndromic true hermaphroditism due to an R-spondin1 (RSPO1) homozygous mutation. Hum Mutat. **2008** Feb; 29(2):220–226. PMID: 18085567.

Tsurusaki Y, Okamoto N, Ohashi H, Mizuno S, Matsumoto N, Makita Y, Fukuda M, Isidor B, Perrier J, Aggarwal S, Dalal AB, Al-Kindy A, Liebelt J, Mowat D, Nakashima M, Saitsu H, Miyake N and Matsumoto N. Coffin–Siris syndrome is a SWI/SNF complex disorder. Clin Genet. **2014** Jun; 85(6):548–554. doi: 10.1111/cge.12225. Epub 2013 Jul 23. PMID: 23815551.

Tsukamoto S, Hara T, Yamamoto A, Ohta Y, Wada A, Ishida Y, Kito S, Nishikawa T, Minami N, Sato K and Kokubo T. Functional analysis of lysosomes during mouse preimplantation embryo development. J Reprod Dev. **2013**; 59(1): 33–39. Epub 2012 Oct 19. PMID: 23080372.

Tucker EJ, Mimaki M, Compton AG, McKenzie M, Ryan MT and Thorburn DR. Next-generation sequencing in molecular diagnosis: NUBPL mutations highlight the challenges of variant detection and interpretation. Hum Mutat. **2012** Feb; 33(2):411–418. doi: 10.1002/humu. 21654. Epub 2011 Dec 22. PMID: 22072591.

Turan S and Bastepe M. The GNAS complex locus and human diseases associated with loss-of-function mutations or epimutations within this imprinted gene. Horm Res Paediatr. **2013**; 80(4):229–241. doi: 10.1159/000355384. Epub 2013 Oct 3. PMID: 24107509.

Turnpenny PD, Ellard S Emery's Elements of Medical Genetics, 12th Revised edition. **2004** Churchill Livingstone.

Tzu J and Marinkovich MP. Bridging structure with function: structural, regulatory, and developmental role of laminins. Int J Biochem Cell Biol. **2008**; 40(2):199–214. Epub 2007 Aug 6. PMID: 17855154.

U

Ungerer M, Knezovich J and Ramsay M. *In utero* alcohol exposure, epigenetic chan-ges, and their consequences. Alcohol Res. **2013**; 35(1):37–46. PMID: 24313163.

V

Vafai SB and Mootha VK. Mitochondrial disorders as windows into an ancient organelle. Nature. **2012** Nov 15; 491(7424):374–383. doi: 10.1038/nature11707. Review. PMID: 23151580.

van Amerongen R and Nusse R. Towards an integrated view of Wnt signaling in development. Development. **2009** Oct; 136(19):3205–3214. doi: 10.1242/dev. 033910.Review. PMID: 19736321.

van de Kamp JM, Lefeber DJ, Ruijter GJ, Steggerda SJ, den Hollander NS, Willems SM, Matthijs G, Poorthuis BJ, Wevers RA. Congenital disorder of glycosylation type Ia presenting with hydrops fetalis. J Med Genet. **2007** Apr; 44(4):277-280. Epub 2006 Dec 8. PMID:17158594.

van Dijk M, Thulluru HK, Mulders J, Michel OJ, Poutsma A, Windhorst S, Kleiverda G, Sie D, Lachmeijer AM and Oudejans CB. HELLP babies link a novel lincRNA to the trophoblast cell cycle. J Clin Invest. **2012** Nov 1; 122(11): 4003–4011. doi: 10.1172/JCI65171. Epub 2012 Oct 24. PMID: 23093777.

van Ooij C, Snyder RC, Paeper BW and Duester G. Temporal expression of the human alcohol dehydrogenase gene family during liver development correlates with differential promoter activation by hepatocyte nuclear factor 1, CCAAT/enhancer-binding protein alpha, liver activator protein, and D-element-binding protein. Mol Cell Biol. **1992** Jul; 12(7):3023–3031. PMID: 1620113.

Vaquerizas JM, Kummerfeld SK, Teichmann SA and Luscombe NM. A census of human transcription factors: function, expression and evolution. Nat Rev Genet. **2009** Apr; 10(4):252–263. doi: 10.1038/nrg2538.

Varelas X and Wrana JL. Coordinating developmental signaling: novel roles for the Hippo pathway. Trends Cell Biol. **2012** Feb; 22(2):88–96. doi: 10.1016/j. tcb.2011.10.002. Epub 2011 Dec 5. PMID: 22153608.

Varjosalo M and Taipale J. Hedgehog: functions and mechanisms. Genes Dev. **2008** Sep 15; 22(18):2454–2472. doi: 10.1101/gad.1693608. Review. PMID: 18794343.

Vivante A, Kohl S, Hwang DY, Dworschak GC and Hildebrandt F. Single-gene causes of congenital anomalies of the kidney and urinary tract (CAKUT) in humans. Pediatr Nephrol. **2014** Apr; 29(4):695–704. doi: 10.1007/s00467-013-2684-4. Epub 2014 Jan 8. PMID: 24398540.

Vogt MC, Paeger L, Hess S, Steculorum SM, Awazawa M, Hampel B, Neupert S, Nicholls HT, Mauer J, Hausen AC, Predel R, Kloppenburg P, Horvath TL and Brüning JC. Neonatal insulin action impairs hypothalamic neurocircuit formation in response to maternal high-fat feeding. Cell. **2014** Jan 30; 156(3):495–509. doi: 10.1016/j.cell.2014.01.008. Epub 2014 Jan 23. PMID: 24462248.

W

Wagner M and Siddiqui MA. Signal transduction in early heart development (I): cardiogenic induction and heart tube formation. Exp Biol Med (Maywood). **2007** Jul; 232(7):852–865. Review. PMID: 17609501.

Wallenstein MB, Shaw GM, Yang W and Carmichael SL. Periconceptional nutrient intakes and risks of orofacial clefts in California. Pediatr Res. **2013** Oct; 74(4):457–65. doi: 10.1038/pr.2013.115. Epub 2013 Jul 3. PMID: 23823175.

Wanders RJ. Metabolic functions ofperoxisomesin health and disease. Biochimie. **2014** Mar; 98:36–44. doi: 10.1016/j. biochi.2013.08.022. Epub 2013 Sep 3. PMID: 24012550.

Wang ET, Sandberg R, Luo S, Khrebtukova I, Zhang L, Mayr C, Kingsmore SF, Schroth GP and Burge CB. Alternative isoform regulation in human tissue transcriptomes. Nature. **2008** Nov 27; 456(7221):470–476. doi: 10.1038/nature07509. PMID: 18978772.

Wang G and Bieberich E. Prenatal alcohol exposure triggers ceramide-induced apoptosis in neural crest-derived tissues concurrent with defective cranial development. Cell Death Dis. **2010** May 27; 1:e46. doi: 10.1038/cddis. 2010. 22. PMID: 21364652.

Wang S, Bates J, Li X, Schanz S, Ch and ler-Militello D, Levine C, Maherali N, Studer L, Hochedlinger K, Windrem M and Goldman SA. Human iPSC-derived oligodendrocyte progenitorcellscan myelinate and rescue a mouse model of congenital hypomyelination. Cell Stem Cell. **2013** Feb 7; 12(2): 252–264. doi: 10.1016/j.stem.2012.12.002. PMID: 23395447.

Wang X, Cabrera RM, Li Y, Miller DS and Finnell RH. Functional regulation of P-glycoprotein at the blood-brain barrier in proton-coupled folate transporter (PCFT) mutant mice. FASEB J. **2013** Mar; 27(3):1167–1175. doi: 10.1096/fj. 12-218495. Epub 2012 Dec 4.

Waterland RA and Michels KB. Epigenetic epidemiology of the developmental origins hypothesis. Annu Rev Nutr. **2007**; 27:363–388. PMID: 17465856.

Wei R, Yang J, Liu GQ, Gao MJ, Hou WF, Zhang L, Gao HW, Liu Y, Chen GA and Hong TP. Dynamic expression of microRNAs during the differentiation of human embryonic stem cells into insulin-producing cells. Gene. **2013** Apr 15; 518(2):246–255. doi: 10.1016/j.gene.2013.01.038. Epub 2013 Jan 29. PMID: 23370336.

Wieczorek D. Human facial dysostoses. Clin Genet. **2013** Jun; 83(6):499–510. doi: 10.1111/cge.12123. Epub 2013 Apr 8. PMID: 23565775.

Wilhelm M, Schlegl J, Hahne H, Moghaddas Gholami A, Lieberenz M, Savitski MM, Ziegler E, Butzmann L, Gessulat S, Marx H, Mathieson T, Lemeer S, Schnatbaum K, Reimer U, Wenschuh H, Mollenhauer M, Slotta-Huspenina J, Boese JH, Bantscheff M, Gerstmair A, Faerber F and Kuster B. Mass-spectrometry-based draft of the human proteome. Nature. **2014** May 29; 509(7502):582–587. doi: 10.1038/nature13319.

Wilkie AO and Morriss-Kay GM. Genetics of craniofacial development and malformation. Nat Rev Genet. **2001** Jun; 2(6):458–468. Review. PMID: 11389462.

Williams SR, Aldred MA, Der Kaloustian VM, Halal F, Gowans G, McLeod DR, Zondag S, Toriello HV, Magenis RE and Elsea SH. Haploinsufficiency of HDAC4 causes brachydactyly mental retardation syndrome, with brachydactyly type E, developmental delays, and behavioral problems. Am J Hum Genet. **2010** Aug 13; 87(2):219–228. doi: 10.1016/j.ajhg.2010.07.011. PMID: 20691407.

Wingender E, Schoeps T and Dönitz J. TFClass: an expandable hierarchical classification of human transcription factors. Nucleic Acids Res. **2013** Jan; 41(Database issue):D165–D170. doi: 10.1093/nar/gks1123. Epub 2012 Nov 24. PMID: 23180794.

Wolf AR and Mootha VK. Functional genomic analysis of human mitochondrial RNA processing. Cell Rep. **2014** May 8; 7(3):918–931. doi: 10.1016/j.celrep.2014.03.035. Epub 2014 Apr 18. PMID: 24746820.

Wonders CP, Taylor L, Welagen J, Mbata IC, Xiang JZ and Anderson SA. A spatial bias for the origins of interneuron subgroups within the medial ganglionic eminence. Dev Biol. **2008** Feb 1; 314(1):127–136. Epub 2007 Nov 28. PMID: 18155689.

Woods NB, Parker AS, Moraghebi R, Lutz MK, Firth AL, Brennand KJ, Berggren WT, Raya A, Izpisúa Belmonte JC, Gage FH and Verma IM. Brief report: efficient generation of hematopoietic precursors and progenitors from human pluripotent stem cell lines. Stem Cells. **2011** Jul; 29(7):1158–1164. doi: 10.1002/stem. 657. PMID: 21544903.

X

Y

Yamazawa K, Ogata T and Ferguson-Smith AC. Uniparental disomy and human disease: an overview. Am J Med Genet C Semin Med Genet. **2010** Aug 15; 154C(3):329–334. doi: 10.1002/ajmg.c.30270. Review. PMID: 20803655.

Yang YJ, Baltus AE, Mathew RS, Murphy EA, Evrony GD, Gonzalez DM, Wang EP, Marshall-Walker CA, Barry BJ, Murn J, Tatarakis A, Mahajan MA, Samuels HH, Shi Y, Golden JA, Mahajnah M, Shenhav R, Walsh CA. Microcephaly gene links trithorax and REST/NRSF to control neural stem cell proliferation and differentiation. Cell. **2012** Nov 21; 151(5):1097–1112. doi: 10.1016/j.cell.2012.10.043. PMID:23178126.

Yasin SA, Ali AM, Tata M, Picker SR, Anderson GW, Latimer-Bowman E, Nicholson SL, Harkness W, Cross JH, Paine SM and Jacques TS. mTOR-dependent abnormalities in autophagy characterize human malformations of cortical development: evidence from focal cortical dysplasia and tuberous sclerosis. Acta Neuropathol. **2013** Aug; 126(2):207–218. doi: 10.1007/s00401-013-1135-4. Epub 2013 Jun 2. PMID: 23728790.

Young JM, Whiddon JL, Yao Z, Kasinathan B, Snider L, Geng LN, Balog J, Tawil R, van der Maarel SM and Tapscott SJ. DUX4 binding to retroelements creates promoters that are active in FSHD muscle and testis. PLoS Genet. **2013** Nov; 9(11):e1003947. doi: 10.1371/journal. pgen.1003947. Epub 2013 Nov 21. PMID: 24278031.

Yashiro K, Riday TT, Condon KH, Roberts AC, Bernardo DR, Prakash R, Weinberg RJ, Ehlers MD and Philpot BD. Ube3a is required for experience-dependent maturation of the neocortex. Nat Neurosci. **2009** Jun; 12(6):777–783. doi: 10.1038/nn. 2327. Epub 2009 May 10. PMID: 19430469.

Young RA. Control of the embryonic stem cell state. Cell. **2011** Mar 18; 144(6): 940–954. doi: 10.1016/j. cell.2011.01.032. Review. PMID: 21414485.

Yurchenco PD and Patton BL. Developmental and pathogenic mechanisms of basement membrane assembly. Curr Pharm Des. **2009**; 15(12):1277–1294. PMID: 19355968.

Z

Zaidi S, Choi M, Wakimoto H, Ma L, Jiang J, Overton JD, Romano-Adesman A, Bjornson RD, Breitbart RE, Brown KK, Carriero NJ, Cheung YH, Deanfield J, DePalma S, Fakhro KA, Glessner J, Hakonarson H, Italia MJ, Kaltman JR,

Kaski J, Kim R, Kline JK, Lee T, Leipzig J, Lopez A, Mane SM, Mitchell LE, Newburger JW, Parfenov M, Pe'er I, Porter G, Roberts AE, Sachidanandam R, Sanders SJ, Seiden HS, State MW, Subramanian S, Tikhonova IR, Wang W, Warburton D, White PS, Williams IA, Zhao H, Seidman JG, Brueckner M, Chung WK, Gelb BD, Goldmuntz E, Seidman CE and Lifton RP. *De novo* mutations in histone-modifying genes in congenital heart disease. Nature. **2013** Jun 13; 498(7453):220–223. doi: 10.1038/nature12141. Epub 2013 May 12. PMID: 23665959.

Zhang L, Li H, Yu J, Cao J, Chen H, Zhao H, Zhao J, Yao Y, Cheng H, Wang L, Zhou R, Yao Z and Guo X. Ectodermal Wnt signaling regulates abdominal myogenesis during ventral body wall development. Dev Biol. **2014** Mar 1; 387(1):64–72. doi: 10.1016/j.ydbio.2013.12.027. Epub 2014 Jan 3. PMID: 24394376.

Zhao R and Goldman ID. The proton-coupled folate transporter: physiological and pharmacological roles. Curr Opin Pharmacol. **2013** Dec; 13(6):875–880. PMID: 24383099.

Zhu X, Wang J, Ju BG and Rosenfeld MG. Signaling and epigenetic regulation of pituitary development. Curr Opin Cell Biol. **2007** Dec; 19(6):605–611. Epub 2007 Nov 7. Review. PMID: 1798885.

Zimmermann MB, Jooste PL and P and av CS. Iodine-deficiencydisorders. Lancet. **2008** Oct 4;372(9645):1251–1262. doi: 10.1016/S0140-6736(08)61005-3. PMID: 18676011.

Zuin J, Franke V, van Ijcken WF, van der Sloot A, Krantz ID, van der Reijden MI, Nakato R, Lenhard B and Wendt KS. A cohesin-independent role for NIPBL at promoters provides insights in CdLS. PLoS Genet. **2014** Feb 13; 10(2):e1004153. doi:10.1371/journal. pgen. 1004153. eCollection 2014 Feb. PMID: 24550742.

Zurita O, Silva Neiva L, Sasarman F and Shoubridge EA. The arginine methyl-transferase NDUFAF7 is essential for complex I assembly and early vertebrate embryogenesis. Hum Mol Genet. 2014May 16;239. [Epub ahead of print]. PMID: 24838397.

INDEX